Log-Linear Models, Extensions, and Applications

Neural Information Processing series

Michael I. Jordan and Thomas Diettrich, Editors

A complete list of books published in the Neural Information Processing series appears at the back of this book.

Log-Linear Models, Extensions, and Applications

Editors:

Aleksandr Aravkin
IBM T.J. Watson Research Center
Yorktown Heights, NY 10589

saravkin@us.ibm.com

Anna Choromanska
Courant Institute, NYU
New York, NY 10012

achoroma@cims.nyu.edu

Li Deng
Microsoft Research
Redmond, WA 98052

deng@microsoft.com

Georg Heigold
Google Research
Mountain View, CA 94043

heigold@google.com

Tony Jebara
Columbia University
New York, NY 10027

jebara@cs.columbia.edu

Dimitri Kanevsky
Google Research
New York, NY 10011

dkanevsky@google.com

Stephen J. Wright
University of Wisconsin
Madison, WI 53706

swright@cs.uwisc.edu

The MIT Press
Cambridge, Massachusetts
London, England

© 2018 Massachusetts Institute of Technology

All rights reserved. No part of this book may be reproduced in any form by any electronic or mechanical means (including photocopying, recording, or information storage and retrieval) without permission in writing from the publisher.

This book was set in LATEX by author. Printed and bound in the United States of America.

Library of Congress Cataloging-in-Publication Data is available

ISBN: 978-0-262-03950-5

10   9   8   7   6   5   4   3   2   1

Contents

1   Some Insights into the Geometry and Training of Neural Networks

Ewout van den Berg   evandenberg@us.ibm.com
IBM Watson
Yorktown Heights, NY, USA

Neural networks have been successfully used for classification tasks in a rapidly growing number of practical applications. Despite their popularity and widespread use, there are still many aspects of training and classification that are not well understood. In this work we aim to provide some new insights into training and classification by analyzing neural networks from a feature-space perspective. We review and explain the formation of decision regions and study some of their combinatorial aspects. We place a particular emphasis on the connections between the neural network weight and bias terms and properties of decision boundaries and regions that exhibit varying levels of classification confidence. We show how the error backpropagates in these regions and emphasize the important role they have in the formation and localization of gradients. The connection between scaling of the weight parameters and the localization of gradients, in combination with the density of the training samples, helps shed more light on the vanishing gradient problem and explains the need for regularization. In addition, it suggests an approach for subsampling training data to improve performance.

1.1   Introduction

Neural networks have been successfully used for classification tasks in applications such as pattern recognition (2), speech recognition (10), and numerous others (25). Despite their widespread use, the understanding of neural networks is still incomplete, and they often remain treated as black boxes. In this work we provide new insights into training and classification by analyzing neural networks from a feature-space perspective. We consider feedforward neural networks in which input vectors $x_0 \in \mathbb{R}^d$ are propagated through n successive layers, each of the form

$$x_k = \nu_k(A_k x_{k-1} - b_k), \tag{1.1}$$

where ν_k is a nonlinear activation function that acts on an affine transformation of the output x_{k-1} from the previous layer, with weight matrix A_k and bias vector b_k. Neural networks are often represented as graphs and the entries in vectors x_k are therefore often referred to as nodes or units. There are three main design parameters in a feedforward neural network architecture: the number of layers or depth the network, the number of nodes in each layer, and the choice of activation function. Once these are fixed, neural networks are training by adjusting only the weight and bias terms.

Although most of the results and principles in this chapter apply more generally, we predominantly consider neural networks with sigmoidal activation functions that are convex-concave and differentiable. To keep the discussion concrete we focus on a symmetrized version of the logistic function that acts elementwise on its input as

$$\sigma_\gamma(x) = 2\ell_\gamma(x) - 1, \qquad \text{with} \qquad \ell_\gamma(x) = \frac{1}{1 + e^{-\gamma x}}. \tag{1.2}$$

This function can be seen as a generalization of the hyperbolic tangent, with $\sigma_\gamma(x) = \tanh(\gamma x/2)$. We omit the subscript γ when $\gamma = 1$, or when its exact value does not matter. For simplicity, and with some abuse of terminology we refer to σ_γ as the sigmoid function, irrespective of the value of γ, and use the term logistic function for ℓ_γ. Examples of several instances of σ_γ and their first-order derivatives are plotted in Figure 1.1.

The activation function in the last layer has the special purpose of ensuring that the output of the neural network has a meaningful interpretation. The softmax function is widely used and generates an output vector whose entries are defined as

$$[\mu(x)]_i = \frac{e^{x[i]}}{\sum_{j=1}^{k} e^{x[j]}}. \tag{1.3}$$

Exponentiation and normalization ensures that all output values are non-negative and sum up to one, and the output of node i can therefore be interpreted as an estimate of the posterior probability $p(\text{class} = i \mid x)$. That is, we can define the estimated probabilities as $\hat{p}_s(\text{class} = i \mid x) := [x_n(x)]_i$, where s is a vector containing the network weight and bias parameters, and

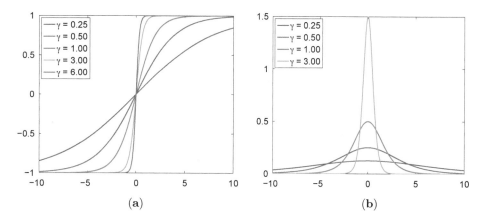

Figure 1.1: Different instances of (a) the hyperbolic tangent function σ_γ and (b) the derivatives.

$x_n(x)$ is the output at layer n corresponding to input $x_0 = x$. The network parameters s are typically learned from domain-specific training data. In supervised training for multiclass classification this training data comes in the form of a set of tuples $\mathcal{T} = \{(x, c)\}$, each consisting of a sample feature vector $x \in \mathbb{R}^d$ and its associated class label c. Training the network is done by minimizing a suitable loss function. Throughout this work we consider

$$\underset{s}{\text{minimize}} \quad \phi(s) := \frac{1}{|\mathcal{T}|} \sum_{(x,c) \in \mathcal{T}} f(s; x, c), \tag{1.4}$$

with the cross-entropy function

$$f(s; x, c) = -\log \hat{p}_s(c \mid x).$$

We denote the class c corresponding to feature vector $x \in \mathbb{R}^d$ as $c(x)$, which, in practice, is known only for all points in the training set. For notational convenience we also write $f(x)$ to mean $f(s; x, c(x))$. The loss function $\phi(s)$ is highly nonconvex in s making (1.4) particularly challenging to solve. However, even if it could be solved, care needs to be taken not to overfit the data to ensure that the network generalizes to unseen data. This can be achieved, for example, through regularization, early termination, or by limiting the model capacity of the network.

The outline of the chapter is as follows. In Section 1.2 we review the definition of halfspaces and the formation of decision regions. In Section 1.3 we look at combinatorial properties of the decision regions, their ability to separate or approximate different classes, and possible generalizations. Section 1.4 analyzes the connection between the decision regions and the gradient with respect to the different network parameters. Topics related to the training of neural networks including backpropagation, regularization, the contribution of individual training samples to the gradient, and importance sampling are discussed in Section 1.5. We conclude with a discussion and future work in Section 1.6.

Throughout the chapter we use the following notational conventions. Matrices are indicated by capitals, such as A for the weight matrices; vectors are denoted by lower-case roman letters. Sets are denoted by calligraphic capitals. Subscripted square brackets denote indexing, with $[x]_j$, $[A]_i$, and $[A]_{i,j}$ denoting respectively the j-th entry of vector x, the i-th row of A as a column vector, and the (i, j)-th entry of matrix A. Square brackets are also used to denote vector or matrix instances with commas separating entries within one row, and semicolon separating rows in in-line notation. We denote vectors of all ones by e (when raised to a power, e^x always represents the natural exponent). The largest singular value of A is denoted $\sigma_{\max}(A)$, from the context it will be clear that this is not a particular instance of the sigmoidal function σ_γ. The vector ℓ_1, and ℓ_2 norms refer to the one- and two norms; that is, the sum of absolute values and the Euclidean norm, respectively. There should be no confusion between this and the ℓ_γ logistic function.

1.2   Formation of Decision Regions

Decision regions can be described as those regions or sets of points in the feature space that are classified as a certain class. Classification in neural networks is soft in the sense that it comes as a vector of posterior probabilities rather than a hard assignment to one class or another. When talking about decision regions for a class c we could consider the set of points where the posterior probability is highest among the classes:

$$\mathcal{C}_c := \{x \in \mathbb{R}^d \mid \hat{p}_s(c \mid x) = \max_j \hat{p}_s(j \mid x)\},$$

or exceeds a given threshold:

$$\mathcal{C}_c := \{x \in \mathbb{R}^d \mid \hat{p}_s(c \mid x) \geq \tau\}. \tag{1.5}$$

Although this section discusses the formation and role of decision regions and its boundaries, we will not use any formal definition of decision regions. However, the intuitive notion used closely follows definition (1.5).

As we will see in this section, decision regions are formed as input is propagated through the network. Even though the form (1.1) of all the layers is identical, we can nevertheless identify two distinct stages in region formation. The first stage defines a collection of halfspaces and takes place in the first layer of the network. The second stage takes place over the remaining layers in which intermediate regions are successively combined to form the final decision regions, starting with the initial set of halfspaces. The generation of halfspaces or hyperplanes in the first layer of the neural network and their combination in subsequent layers is well known (see for example (2, 16)). The formation of soft decision boundaries and some of their properties does not appear to have been studied widely. We now take a closer look at the two stages, with a particular emphasis on the role of the sigmoidal activation function in each of these stages.

1.2.1 Definition of halfspaces

The output of the first layer in the network can be written as $y = \sigma(Ax - b)$ with $x \in \mathbb{R}^d$. For an individual unit j this reduces to $\sigma(\langle a, x \rangle - \beta)$, where $a \in \mathbb{R}^d$ corresponds to $[A]_j$, the j-th row of A, and $\beta = [b]_j$. When applied over all points of the feature space, the affine mapping $\langle a, x \rangle - \beta$ generates a linear gradient, as shown in Figure 1.2(a). The output of a unit is then obtained by applying the sigmoid function to these intermediate values. When doing so, assuming throughout that $a \neq 0$, two prominent regions form: one with values close to -1 and one with values close to $+1$. In between the two regions there is a smooth transition region, as illustrated in Figure 1.2(b). The center of the transition region consists of all feature points whose output value equal zero. It can be verified that this set is given by all points $x = \beta a/\|a\|_2^2 + v$ such that $\langle a, v \rangle = 0$, and therefore describes a hyperplane. The normal direction of the hyperplane is given by a, and the exact location of the hyperplane is determined by a shift along this normal, controlled by both β and $\|a\|_2$. The region of all points that map to nonnegative values forms a halfspace, and because the linear functions can be chosen independently for each unit, we can define as many halfspaces as there are units in the first layer. As the transition between the regions on either side of the hyperplane is gradual it is convenient to work with soft boundaries and interpret the output values as a confidence level of set membership with values close to $+1$ indicating strong membership, those close to -1 indicating strong non-membership, and with decreasing confidence levels in between. For simplicity we use the term halfspace for both the soft and sharp versions of the region.

In addition to normal direction and location, halfspaces are characterized by the sharpness of the transition region. This property can be controlled in two similar ways (see also Section 1.5.2). The first is to scale both a and β by a positive scalar γ. Doing so does not affect the location or orientation of the

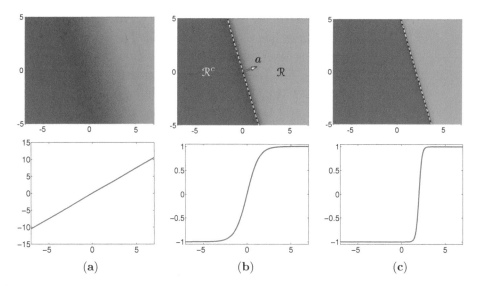

Figure 1.2: The mapping of points x in the feature space using (a) the linear transformation $\langle a, x \rangle - \beta$ with $a = [1.5, 0.5]^T$ and $\beta = 0$; (b) the nonlinearity $\sigma(\langle a, x \rangle - \beta)$ with the same values for a and β; and (c) the nonlinearity $\sigma(\langle a', x \rangle - \beta)$ with $a' = 4a$ and $\beta = 12$. Shown are the output values for points in a rectangular region of the feature space (top row), and for points x with $[x]_2 = 0$ (bottom row).

hyperplane but does scale the input to the sigmoid by the same quantity. As a consequence, choosing $\gamma > 1$ shrinks the transition region, whereas choosing $\gamma < 1$ causes it to widen. The second way is to replace the activation function σ by σ_γ. Scaling only a affects the sharpness of the transition in the same way, but also results in a shift of the hyperplane along the normal direction whenever $\beta \neq 0$. Note however that the activation functions are typically fixed and the properties of the halfspaces are therefore controlled only by the weight and bias terms. Figure 1.2(c) illustrates the sharpening of the halfspace and the use of β to change its location.

1.2.2 Combination of intermediate regions

The second layer combines the halfspace regions defined in the first layer resulting in new regions in each of the output nodes. In case of step-function activations, the region combinations are similar to set operations including complements (^{c}), intersection ($\cap$), and unions ($\cup$). The same operations are used in subsequent layers, thereby enabling the formation of increasingly complex regions. The use of a sigmoidal function instead of the step function does not significantly change the types of operations, although some care needs to be taken. Some operations are best explained when working with input coming from the logistic function (with values ranging from 0 to 1) rather than from the sigmoid function[1] (ranging from -1 to 1).

1.2.2.1 Elementary Boolean operations

To make the operations discussed in this section more concrete we apply them to input generated by a first layer with the following parameters:

$$A_1 = \begin{bmatrix} 9 & 1 \\ -2 & 6 \end{bmatrix}, \quad b_1 = \begin{bmatrix} -2 \\ -1 \end{bmatrix}, \quad \text{and} \quad \nu_1 = \sigma_3.$$

The two resulting halfspace regions $\mathcal{R}_1$ and $\mathcal{R}_2$ are illustrated in Figures 1.3(a) and 1.3(b). For simplicity we denote the parameters for the second layer by A and b, omitting the subscripts. In addition, we omit all entries that are not relevant to the operation and apply appropriate padding with zeros where needed.

Constants Constants can be generated by choosing $A = 0$ and choosing a sufficiently large positive or negative offset values. For example, choosing $b = 100$ gives a region that spans the entire domain (representing the logical TRUE), whereas choosing $b = -100$ results in the empty set (or logical FALSE).

Unary operations The simplest unary operation, the (Boolean) identity function, can be defined as

$$A_I = 1, \qquad b_I = 0. \tag{Identity}$$

This function works well when used in conjunction with a step function, but has an undesirable damping effect when used with the sigmoid function: input values up to 1 are mapped to output values up to $\sigma(1) \approx 0.46$, and likewise for negative values. While such scaling may be desirable in certain cases, we would like to preserve the clear distinction between high and low confidence regions. We can do this by scaling up A, which amplifies the input to the sigmoid function and therefore its output. Choosing $A_I = 3$, for example, would increase the maximum confidence level to $\sigma(3) \approx 0.91$.

---

1. Note however that output x from a sigmoid function can easily be mapped to the output $x' = (x-1)/2$ from a logistic function, and vice versa, by appropriately scaling the weight and bias terms in the next layer. Any linear operation $Ax' - b$ on the logistic output then becomes $A(x-1)/2 - b = \tilde{A}x - \tilde{b}$ with $\tilde{A} = A/2$ and $\tilde{b} = b + Ae/2$. In other words, with appropriate changes in A and b we can always choose which of the two activation functions the input comes from, regardless of which function was actually used.

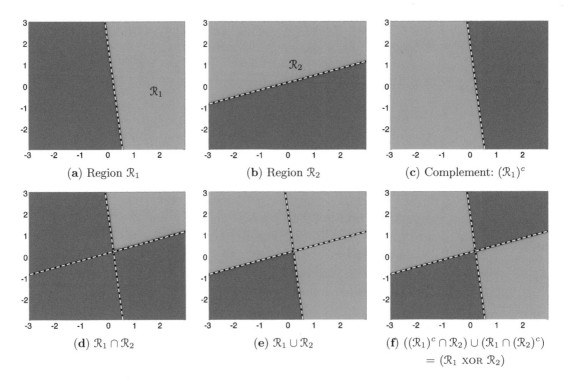

Figure 1.3: Illustration of (a), (b) regions defined by the first neural network layer, and (c)–(f) various Boolean set operations applied to them in subsequent layers.

As noted towards the end of Section 1.2.1, the same can be achieved by working with the activation function σ_3, and to avoid getting distracted by scaling issues like these we will work with $\nu_2 = \sigma_3$ throughout this section. We note that the identity function can be approximated very well by scaling down the input and taking advantage of the near-linear part of the sigmoid function around zero. The output can then be scaled up again in the next layer to achieve the desired result. Similar to the identity function, we define the complement of a set as

$$A_c = -1, \qquad b_c = 0. \tag{Complement}$$

The application of this operator to $\mathcal{R}_1$ is illustrated in Figure 1.3(c). Just to be clear, note that in this case the full parameters to the second layer would be $A = [-1, 0]$ and $b = 0$.

Binary operations When taking the intersection of regions $\mathcal{R}_1$ and $\mathcal{R}_2$ we require that the output values of the corresponding units in the network sum up to a value close to two. This is equivalent to saying that when we subtract a relatively high value, say 1.5, from the sum, the outcome should remain positive. This suggests the following parameters for binary intersection:

$$A_\cap = [1, 1], \qquad b_\cap = 1.5. \qquad\qquad\qquad \text{(Intersection)}$$

We now combine the intersection and complement operations to derive the union of two sets, and to illustrate how complements of sets can be applied during computations. By De Morgan's law, the union operator can be written as $\mathcal{R}_1 \cup \mathcal{R}_2 = ((\mathcal{R}_1)^c \cap (\mathcal{R}_2)^c)^c$. Evaluation of this expression is done in three steps: taking the individual complements of $\mathcal{R}_1$ and $\mathcal{R}_2$, applying the intersection, and taking the complement of the result. This can be written in linear form as

$$A_c \left(A_\cap \left(\begin{bmatrix} A_I & \\ & A_c \end{bmatrix} x + \begin{bmatrix} b_I \\ b_c \end{bmatrix} \right) + b_\cap \right) + b_c.$$

Substituting the weight and bias terms and simplifying yields parameters for the union:

$$A_\cup = [1, 1], \qquad b_\cup = -1.5. \qquad\qquad\qquad \text{(Union)}$$

It can be verified that the intersection can similarly be derived from the union operator based using $\mathcal{R}_1 \cap \mathcal{R}_2 = ((\mathcal{R}_1)^c \cup (\mathcal{R}_2)^c)^c$ Results obtained with both operators are shown in Figures 1.3(e) and 1.3(f).

1.2.2.2 General n-ary operations

We now consider general operations that combine regions from more than two units. It suffices to look at a single output unit $\sigma(\langle a, x \rangle - \beta)$ with weight vector a and bias term β. Any negative entry in a means that the corresponding input region is negated and that its complement is used, while zero valued entries indicate that the corresponding region is not used. Without loss of generality we assume that all input regions are used and normalized such that all entries in a can be taken strictly positive. We again start by looking at the idealized situation where inputs are generated using a step function with outputs -1 or 1. When a is the vector of all ones, and k out of n inputs are positive we have $\langle a, x \rangle = k - (n - k) = 2k - n$. Choosing activation level $\beta = 2k - n - 1$ therefore ensures that the output of the unit is positive whenever at least k out of n inputs are positive. As extreme cases of this we obtain the n-ary intersection with $k = n$, and the n-ary union by choosing $k = 1$. Weights can be adjusted to indicate how many times each region gets counted.

It was noted by Huang and Littmann (13) that complicated and highly non-intuitive regions can be formed with the general n-ary operations, even in the second layer. As an example, consider the eight halfspace boundaries

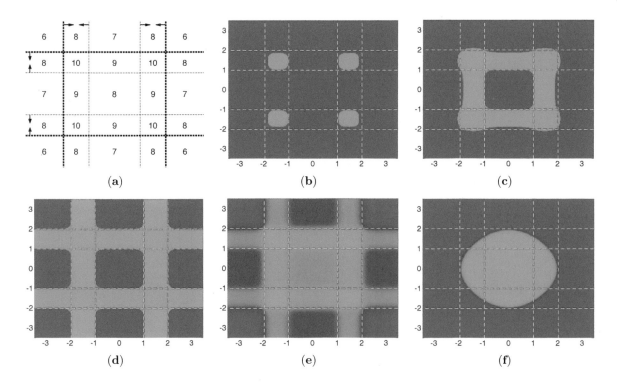

Figure 1.4: Application of the *n*-ary operator. Plot (a) shows the location and orientation of the eight hyperplanes and the total weight for each of the cells when using a weight of two for the outer hyperplanes (thick dashed line) and unit weight for the inner hyperplanes (thin dashed line). Plots (b) and (c) show the regions formed when choosing $\beta = 9.5$ and $\beta = 8.5$ respectively, with activation function $\nu_1 = \ell_{10}$ and $\nu_2 = \sigma_{50}$. Plot (d) shows the region obtained when assigning unit weights to each hyperplanes and using $\beta = 4.5$. In plot (e) we change the settings from plot (b) by replacing the second activation function to σ_1 and using $\beta = 6.5$. The lack of amplification results in a region with four different confidence levels. Plot (f) illustrates the formation of a smooth circular region using only the outer four hyperplanes together with activation functions $\nu_1 = \ell_1$ and $\nu_2 = \sigma_{50}$, and threshold $\beta = 6.5$.

plotted in Figure 1.4(a). Assume that each halfspace is defined using a unit step function such that points within the halfspace have a value of 1 and the points outside have a value of 0. The intersections of the halfspaces form cells and the value of points within each cell depends on the weights assigned to the defining halfspaces. Figure 1.4(a) shows the value for points in each cell when using a weight of two to the halfspaces bounded by the outer hyperplanes and a unit weight for those bounded by the inner hyperplanes.

Adding up values for so many regions in a single step worsens the scaling issue mentioned for the unitary operator: in this case choosing a threshold of $\beta = 9.5$ leads to values ranging from -3.5 to 0.5 before application of the sigmoid function. Using $\nu_1 = \ell_{10}$ and $\nu_2 = \sigma_{50}$ for amplification with

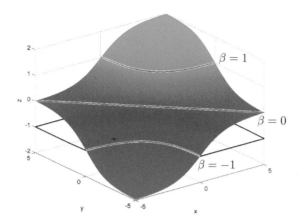

Figure 1.5: Level sets of $z = \sigma(x) + \sigma(y)$ at values $\beta = -1, 0, 1$ along with a slice at $\beta = -1$.

different weight vectors and threshold values we obtain the regions shown in Figures 1.4(b) to 1.4(d).

Removing the large amplification factor in the second layer can lead to regions with low or varying confidence levels. For the mixed weights example, using $\nu_2 = \sigma_1$ and threshold $\beta = 6.5$ causes the intended region to have four distinct confidence levels, as shown in Figure 1.4(e). Low weights can also be leveraged to obtain a parsimonious representation of smooth regions that would otherwise require many more halfspaces. An example of this is shown in Figure 1.4(f) in which the four outer halfspaces with soft boundaries are combined to form a smooth circular region.

1.2.3 Boolean function representation using two layers

As seen from Section 1.2.2.1 neural networks can be used to take the union of intersections of (possibly negated) sets. In Boolean logic this form is called disjunctive normal form (DNF), and it is well known that any Boolean function can be expressed in this form (see also (1)). Likewise we could reverse the order of the union and intersection operators and arrive at conjunctive normal form (CNF), which is equally powerful. Two-layer networks are, in fact, far stronger than this and can be used to approximate general smooth functions. More information on this can be found in (2, Sec. 4.3.2).

1.2.4 Boundary regions and amplification

In the previous subsections we gave a detailed discussion on the formation of decision regions. Here we shift our focus to the boundaries of these regions. The use of sigmoidal nonlinearity functions, rather than step functions, leads to continuous transitions between different regions. The center of the transition regions for a node can be defined as the set of feature points for which the output of that node is zero. Given input x_{k-1} for some node at depth k we first form $\langle a, x_{k-1} \rangle$, then subtract the bias β, and apply the sigmoid function. The output is zero if and only if $\langle a, x_{k-1} \rangle = \beta$, and the transition center therefore corresponds to the level set of $\langle a, x_{k-1} \rangle$ at β. For a fixed a we can thus control the location of the transition by changing β. As an example consider a two-level neural network with the first layer parameterized by $A_1 = I$, $b_1 = 0$, and the second layer by $A_2 = [1, 1]$, $b = \beta$. Writing the input vector as $x_0 = [x, y]$ it can be seen that $A_2 x_1 = \sigma(x) + \sigma(y)$, as illustrated in Figure 1.5. All values greater than β will be mapped to positive values and, as discussed in Section 1.2.2.1, we again see that choosing $\beta > 0$ approximates the intersection of the regions $x \geq 0$ and $y \geq 0$, whereas choosing $\beta < 0$ approximates the union (indicated in the figure by the lines at $z = -1$). What we are interested in here is the location of the transition center. Clearly, making the intersection more stringent by increasing β causes the boundary to shift and the resulting region to become smaller. Another side effect is that the output range of the second layer, which is given by $[\sigma(-2 - \beta), \sigma(2 - \beta)]$, changes. Choosing β close to 2, the supremum of the input signal, means that the supremum of the output is close to zero, whereas the infimum nearly reaches -1. To obtain larger positive confidence levels in the output, without shifting the transition center, we need to amplify the input by scaling A_2 and b_2 by some $\gamma > 1$. In Figure 1.6 we study several aspects of the boundary region corresponding to the setting used for Figure 1.5, with the addition of scaling parameter γ. For a given β we choose γ such that the maximum output of $\sigma(\gamma(2 - \beta))$ is 0.995. Figures 1.6(a)–(c) show the transition region with values ranging from -0.95 and 0.95 along with the center of the transition with value 0 and the region with values exceeding 0.95. Figure 1.6(d) shows the required scaling factors.

The ideal intersection of the two regions coincides with the positive orthant and we define the shift in the transition boundary as the limit of the y-coordinate of the zero crossing as x goes to infinity, giving

$$\lim_{x \to \infty} \sigma^{-1}(\beta - \sigma(x)) = \sigma^{-1}(\beta - 1).$$

The resulting shift values are show in Figure 1.6(e). Another property of interest is the width of the transition region. Similar to the shift we quantify this as the difference between the asymptotic y-coordinates of the -0.95 and 0.95 level set contours as x goes to infinity. We plot the results for several multiples of γ in Figure 1.6(f). As expected, we can see that larger

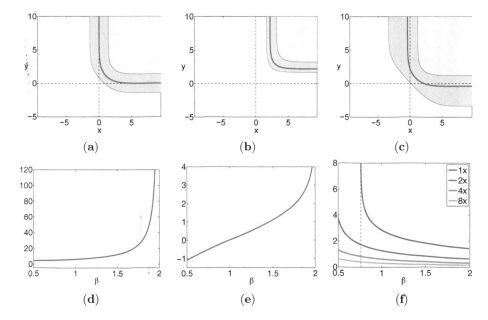

Figure 1.6: Transition regions for $\sigma_\gamma(\sigma(x) + \sigma(y) - \beta)$ with contour lines at 0 and 0.95 of the minimum and maximum values for (a) $\beta = 1$, (b) $\beta = 1.8$, and (c) $\beta = 0.6$. The value of scaling factor γ is chosen such that output range reaches at least ± 0.995. Plot (d) shows the required scaling factor as a function of β. The shift in the transition region is plotted in (e). Plot (f) shows the transition width as a function of β, using different multiples of γ.

amplification reduces the size of the transition intervals. The vertical dashed line indicates the critical value of β at which the -0.95 contour becomes diagonal $(y = -x)$ causing the transition width to become infinite[2]. The same phenomenon happens at smaller β when the multiplication factor is higher.

1.2.5   Continuous to discrete

It can be seen from Figure 1.5 that decision boundaries are traced by level-sets of the input. The level is determined by the bias as well as the shape of the nonlinearity, and can be chosen to generate decision boundaries that look very different from any of those of the input. One example of this was shown in Figure 1.4(f) in which a circular region was generated by four axis-aligned hyperplanes. We now describe another; consider the two hyperplanes in Figures 1.7(a,b), generated in the first layer with respectively $a = [0.1, -0.1]$, $b = 0$, and $a = [0.1, 0.1]$, $b = 0$. The small weights and the

---

2. Note that this breakdown is due only to the definition of the transition width; the transition region itself remains perfectly well defined throughout.

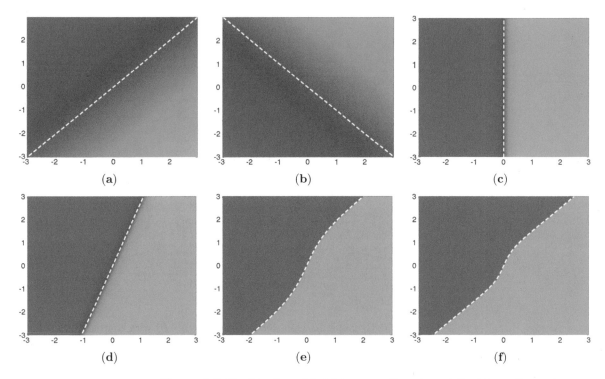

Figure 1.7: Combination of (a,b) two smooth regions defined by a diagonal hyperplane into (c) a vertical region, and (d) a region at an angle of 70 degrees. Using the setup for (d) with scaled weights in the first layer gives the region shown in (e) for weight factor 10, and (f) for weight factor 20.

limited domain size cause the input values to the nonlinearity to be small. As a result, the sigmoid operates in its near-linear region around the origin and therefore resembles scalar multiplication. Consequently, because the normals of the first layers form a basis, we can use the second layer to approximate any operation that would normally occur in the first layer. For example we can choose $a_2 = [\cos(\alpha + \pi/4), \sin(\alpha + \pi/4)]$ and $b_2 = 0$ to generate a close approximation of a hyperplane at angle α (up to a scaling factor this weight matrix is formed by multiplying the desired normal vector a by the rotation on the inverse of the weight matrix of the first layer). The resulting regions of the second layer are shown in Figures 1.7(c,d) for $\alpha = 90°$ and $\alpha = 70°$, respectively. This illustrates that, although somewhat contrived, it is technically possible, at least locally, to change hyperplane orientation after the first layer.

As decision regions propagate and form through one or more layers with modest or large weights, their boundaries become sharper and we see a gradual transition from continuous to discrete network behavior. In the continuous regime, where the transitions are still gradual, the decision boundaries emerge as level sets of slowly varying smooth functions and therefore change continuously and considerably with the choice of bias term. As the boundary regions become sharper the functions tend to piecewise constant causing the level sets to change abruptly only at several critical values while remaining fairly constant otherwise, thus giving more discrete behavior. In Figures 1.7(e,f) we show intermediate stages in which we scale the weights in the first layer Figures 1.7(d) by a factor of 10 and 20, respectively. In addition, it can be seen that scaling in this case does not just sharpen the boundaries, but actually severely distorts them. Finally, it can be seen that the resulting region becomes increasingly diagonal (similar to its sharpened input) as the weights increase. This again emphasizes the more discrete nature of region combinations once the boundaries of the underlying regions are sharp.

1.2.6   Generalized functions for the first layer

The nodes in the first layer define geometric primitives, which are combined in subsequent layers. Depending on the domain it may be desirable to work with primitives other than halfspaces, or to provide a set of different types. This can be achieved by replacing the inner products in the first layer by more general functions $f_\theta(x)$ with training examples x and (possibly shared) parameters θ. The traditional hyperplane is given by

$$f_\theta(x) = \langle a, x \rangle + \beta, \qquad \theta = (a, \beta)$$

For ellipsoidal regions we could instead use

$$f_\theta(x) = \alpha \|Ax - b\|_2^2 + \beta, \qquad \theta = (A, b, \alpha, \beta).$$

More generally, it is possible to redefine the entire unit by replacing both the inner-product and the nonlinearity with a general function to obtain, for example, a radial-basis function unit (15). In Figure 1.8 we illustrate how a mixture of two types of geometric primitives can form regions that cannot be expressed concisely with either type alone.

1.3   Region Properties and Approximation

The hyperplanes defined by the first layer of the neural network partition the space into different regions. In this section we discuss several combinatorial and approximation theoretic properties of these regions. In particular, it shows what the size of the first layer should be in order to represent or approximate certain domains, as the size corresponds directly to the number of hyperplanes.

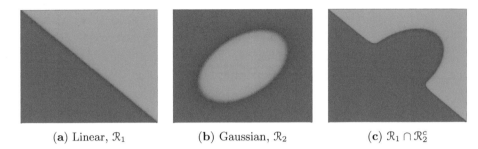

(a) Linear, $\mathcal{R}_1$ (b) Gaussian, $\mathcal{R}_2$ (c) $\mathcal{R}_1 \cap \mathcal{R}_2^c$

Figure 1.8: Shape primitives of type (a) halfspace, and (b) Gaussian, are combined to obtain (c).

1.3.1   Number of regions

One of the most fundamental properties to consider is the maximum number of regions into which $\mathbb{R}^d$ can be partitioned using n hyperplanes. The exact maximum is well known to be

$$r(n, d) = \sum_{i=0}^{d} \binom{n}{i}, \tag{1.6}$$

and is attained whenever the hyperplanes are in general position (21, p.39). With the hyperplanes in place, the subsequent logic layers in the neural network can be used to identify each of these regions by taking the union of (complements of) halfspaces. Individual regions can then be combined using the union operator.

1.3.2   Approximate representation of classes

1.3.2.1   *Polytope representation*

When the set of points $\mathcal{C} \subset \mathbb{R}^d$ belonging to a class form a bounded convex set, we can approximate it by a polytope $\mathcal{P}$ given by the bounded intersection of a finite number of halfspaces. The accuracy of such an approximation can be expressed as the Hausdorff distance between the two sets, defined as:

$$\rho_H(\mathcal{C}, \mathcal{P}) := \max \left[\sup_{x \in \mathcal{C}} d(x, \mathcal{P}), \; \sup_{x \in \mathcal{P}} d(x, \mathcal{C}) \right],$$

with

$$d(x, \mathcal{S}) := \inf_{y \in \mathcal{S}} \|x - y\|_2.$$

For a given class of convex bodies Σ, denote $\delta_H(\mathcal{C}, \Sigma) := \inf_{\mathcal{V} \in \Sigma} \rho_H(\mathcal{C}, \mathcal{V})$. We are interested in $\delta_H(\mathcal{C}, \Sigma)$ when $\Sigma = \mathfrak{R}^d_{(n)}$, the set of all polytopes in $\mathbb{R}^d$ with at most n facets (i.e., generated by the intersection of up to n halfspaces), and in particular how it behaves as a function of n. The following result obtained independently by (4, 8) is given in (5). For every convex body $\mathcal{U}$ there exists a constant $c(\mathcal{U})$ such that

$$\delta_H(\mathcal{U}, \mathfrak{R}^d_{(n)}) \leq \frac{c(\mathcal{U})}{n^{2/(d-1)}}.$$

More interesting perhaps is a lower bound on the approximation distance, which translates into a lower bound on the number of hyperplanes needed to approximately represent the domain. For the unit ball $\mathcal{B}$ we have:

Theorem 1.1. *Let $\mathcal{B}$ denote the unit ball in $\mathbb{R}^d$. Then for sufficiently large n there exists a constant c_d such that*

$$\delta_H(\mathcal{B}, \mathfrak{R}^d_{(n)}) \geq \frac{c_d}{n^{2/(d-1)}}.$$

Proof. For n large enough there exists a polytope $\mathcal{P} \in \mathfrak{R}^d_{(n)}$ with n facets and $\delta := \delta_H(\mathcal{B}, \mathcal{P}) \leq 9/64$. Each of the n facets in $\mathcal{P}$ is generated by a halfspace, and we can use each halfspace to generate a point on the unit sphere in $\mathbb{R}^d$ such that the surface normal at that point matches the outward normal of the halfspace. We denote the set of these points by $\mathcal{N}$, with $|\mathcal{N}| = n$. Now, take any point x on the unit sphere. From the definition of δ it follows that the maximum distance between x and the closest point on one of the hyperplanes bounding the halfspaces is no greater than δ. From this it can be shown that the distance to the nearest point in $\mathcal{N}$ is no greater than $\epsilon := 2\sqrt{\delta}$. Moreover, because the choice of x was arbitrary, it follows that $\mathcal{N}$ defines an ϵ-net of the unit sphere. Lemma 1.2 below shows that the cardinality $|\mathcal{N}| \geq \frac{c'}{\epsilon^{(d-1)}}$. Substituting $\epsilon = 2\sqrt{\delta}$ gives

$$n \geq \frac{c'}{2^{(d-1)} \delta^{(d-1)/2}}, \quad \text{or} \quad \delta \geq \frac{c_d}{n^{2/(d-1)}}.$$

$\square$

Lemma 1.2. *Let $\mathcal{N}$ be an ϵ-net of the unit sphere $\mathcal{S}^{d-1}$ in $\mathbb{R}^d$ with $\epsilon \leq 3/4$, then*

$$|\mathcal{N}| \geq \sqrt{2\pi(d-1)/d} \cdot \epsilon^{1-d}.$$

Proof. By definition of the ϵ-net, we obtain a cover for $\mathcal{S}^{d-1}$ by placing balls of radius ϵ at all $x \in \mathcal{N}$. The intersection of each ball with the sphere gives a spherical cap. The union of the spherical caps covers the sphere and $|\mathcal{N}|$

times the area of each spherical cap must therefore be at least as large as the area of the sphere. A lower bound on the number of points in $\mathcal{N}$ is therefore obtained by the ratio ν between the area of the sphere and that of the spherical cap (see also (24, Lemma 2)). Denoting by $\varphi = \arccos(1 - \frac{1}{2}\epsilon^2)$ the half-angle of the spherical cap it follows from (14, Corollary 3.2(iii)) that ν satisfies

$$1/\nu < \frac{1}{\sqrt{2\pi(d-1)}} \cdot \frac{1}{\cos\varphi} \cdot \sin^{d-1}\varphi,$$

whenever $\varphi \leq \arccos 1/\sqrt{d}$. This bound can be substituted into the second term above to obtain $\sqrt{d}$, and it can be verified to hold whenever $\epsilon \leq 3/4$. It further holds that $\sin\varphi < \epsilon$ which, after rewriting, gives the desired result.   □

1.3.2.2   *More efficient representations*

From Theorem 1.1 we see that a large number of supporting hyperplanes is needed to define a polytope that closely approximates the unit ℓ_2-norm ball. Approximating such a ball or any other convex sets by the intersection of a number of halfspaces can be considered wasteful, however, since it uses only a single region out of the maximum $r(n, d)$ given by (1.6). This fact was recognized by Cheang and Barron (6), and they proposed an alternative representation for unit balls that only requires $\mathcal{O}(d^2/\delta^2)$ halfspaces—far fewer than the conventional $\mathcal{O}(1/\delta^{(d-1)/2})$. The construction is as follows: given a set of n suitably chosen halfspaces $\mathcal{H}_i$ and the indicator function $1_{\mathcal{H}_i}(x)$ which is one if $x \in \mathcal{H}_i$ and zero otherwise. Typically these halfspaces are used to define polytope $\mathcal{P} := \{x \in \mathbb{R}^d \mid \sum_i 1_{\mathcal{H}_i}(x) = n\}$, that is, the intersection of all halfspaces. The (non-convex) approximation proposed in (6) is of the form

$$\mathcal{Q} := \{x \in \mathbb{R}^d \mid \sum_i 1_{\mathcal{H}_i}(x) \geq k\},$$

which consists of all points that are contained in at least k halfspaces. This representation is shown to provide far more efficient approximations, especially in high dimensions. As described in Section 1.2.2.2, this construction can easily be implemented as a neural network. A similar approximation for the Euclidean ball, which also takes advantage of smooth transition boundaries is shown in Figure 1.4(f).

1.3.3 Bounds on the number of separating hyperplanes

In many cases, it suffices to simply distinguish between the different classes instead of trying to exactly trace out their boundaries. Doing so may reduce the number of parameters and additionally help reduce overfitting. The bound in Section 1.3.1 gives the maximum number of regions that can be separated by a given number of hyperplanes. Classes found in practical applications are extremely unlikely to exactly fit these cells, and we can therefore expect that more hyperplanes are needed to separate them. We now look at the maximum number of hyperplanes that is needed.

1.3.3.1 Convex sets

In this section we assume that the classes are defined by convex sets whose intersection is either empty or of measure zero. We are interested in finding the minimum number of hyperplanes needed such that each pair of classes is separated by at least one of the hyperplanes. In the worst case, a hyperplane is needed between any pair of n classes, giving a maximum of $\binom{n}{2}$ hyperplanes, independent of the ambient dimension. That this maximum can be reached was shown by Tverberg (23) who provides a construction due to K.P. Villanger of a set of n lines in $\mathbb{R}^3$ such that any hyperplane that separates one pair of lines, intersects all others. Here we describe a generalization of this construction for odd dimensions $d \geq 3$.

Theorem 1.3. *Let $A = [A_1, A_2, \ldots, A_n]$ be a full-spark(7) matrix with blocks A_i of size $d \times (d-1)/2$, with odd $d \geq 3$. Let b_i, $i = 1, \ldots, n$ be vectors in $\mathbb{R}^d$ such that $[A_i, A_j, b_i - b_j]$ is full rank for all $i \neq j$. The subspaces*

$$\mathcal{S}_i = \{x \in \mathbb{R}^d \mid x = A_i v + b_i, \ v \in \mathbb{R}^{(d-1)/2}\}.$$

are pairwise disjoint and any hyperplane separating $\mathcal{S}_i$ and $\mathcal{S}_j$, $i \neq j$, intersects all $\mathcal{S}_k$, $k \neq i, j$.

Proof. Any pair of subspaces $\mathcal{S}_i$ and $\mathcal{S}_j$ intersects only if there exist vectors u, v such that

$$A_i u + b_i = A_j v + b_j, \quad \text{or} \quad [A_i, A_j] \begin{bmatrix} u \\ -v \end{bmatrix} = b_j - b_i.$$

It follows from the assumption that $[A_i, A_j, b_j - b_i]$ is full rank, that no such two vectors exist, and therefore that all subspaces are pairwise disjoint.

Any hyperplane $\mathcal{H}_{i,j}$ separating $\mathcal{S}_i$ and $\mathcal{S}_j$ is of the form $a^T x = \beta$. To avoid intersection with $\mathcal{S}_i$ we must have $a^T (A_i v + b) \neq \beta$ for all $v \in \mathbb{R}^{d-1}$, which is satisfied if and only if $a^T A_i = 0$. It follows that we must also have $a^T A_j = 0$, and therefore that a is a normal vector to the $(d-1)$-subspace spanned by $[A_i, A_j]$. From the full-spark assumption on A it follows that $a^T A_k \neq 0$ for all $k \neq i, j$, which shows that $\mathcal{H}_{i,j}$ intersects the corresponding $\mathcal{S}_k$. The result follows since the choice of i and j was arbitrary. $\qquad\square$

Random matrices A and vectors b_i with entries i.i.d. Gaussian satisfy the conditions in Theorem 1.3 with probability one, thereby showing the existence of the desired configurations. A simple extension of the construction to dimension $d+1$ is obtained when generating subspaces $\mathcal{S}'_i \subset \mathbb{R}^{d+1}$ by matrices A'_i, formed by appending a row of zeros to A_i and adding a column corresponding to the last column of the $d \times d$ identify matrix, and vectors $b'_i = [b_i; 0]$. Pach and Tardos (18) further show that the lines in the construction described by Tverberg can be replaced by appropriately chosen unit segments. Adding a sufficiently small ball in the Minkowski sense then results in n bounded convex sets with non-empty interior whose separation requires the maximum $\binom{n}{2}$ hyperplanes.

1.3.3.2   Point sets

When separating a set of n points, the maximum number of hyperplanes needed is easily seen to be $n-1$; we can cut off a single extremal point of subsequent convex hulls until only a single point is left. This maximum can be reached, for example when all points lie on a straight line. For a set of points in general position, it is shown in (3) that the maximum number $f(n, d)$ of hyperplanes needed satisfies

$$\lceil (n-1)/d \rceil \leq f(n, d) \leq \lceil (n - 2^{\lceil \log d \rceil})/d \rceil + \lceil \log d \rceil.$$

Based on this we can expect the number of hyperplanes needed to separate a family of unit balls to be much smaller than the maximum possible $\binom{n}{2}$, whenever $n > d + 1$.

1.3.3.3   Non-convex sets

The interface between two non-convex sets can be arbitrarily complex, which means that there are no meaningful bounds on the number of hyperplanes needed to separate general sets.

1.4 Gradients

Parameters in the neural network are learned by minimizing a loss function over the training set, using for example stochastic gradient descent on the formulation shown in (1.4). The gradient of such a loss function decouples over the training samples and can be written as

$$\nabla\phi(s) = \tfrac{1}{|\mathcal{T}|} \sum_{(x,c)\in\mathcal{T}} \nabla f(s;x,c) \tag{1.7}$$

where each term can be evaluated using backpropagation (20). The idea of the section is to explore how points contribute when they are part of a training set. That is, for a given parameter set s, and with the class information $c = c(x)$ assumed to be known, we are interested in $\nabla f(x)$; the behavior of ∇f as a function of x over the entire feature space. We will see that some points in the training set contribute more to the gradient than others. So, instead of just looking at the total gradient, we look at the contribution to the gradient of each point: points that have a large relative contribution to the gradient can be said to be more informative than those that do not contribute much (the amount of contribution of each point typically changes during optimization). Throughout this and the next section we use the word 'gradient' loosely and also use it to refer to blocks of gradient entries corresponding to the parameters a layer, individual entries, or the gradient field $\nabla f(x)$ of those quantities over the entire feature space. The exact meaning should be clear from the context.

1.4.1 Motivational example

We illustrate the relative importance of different training samples using a simple one-dimensional example. We define a basic two-layer neural network in which the first layer defines a hyperplane $\alpha x = \beta$ with nonlinearity $\nu_1 = \sigma$, and in which the second layer applies the identity function followed by nonlinearity $\nu_2 = \ell_5$ for amplification (for simplicity we look only at one class and use a logistic function instead of the softmax function). Choosing $\alpha = 1$ and $\beta = 0$ defines the region shown in Figure 1.9(a). Now, suppose that all points $x \in [-12, 12]$ belong to the same class and should therefore be part of this region. Intuitively, it can be seen that slight changes in the location of the hyperplane or in the steepness of the transition will have very little effect on the output of the neural network for input points $|x| \geq 5$, say, since values close to one or zero remain so after the perturbation. As such, we expect that in these regions the gradient with respect to α and β will be

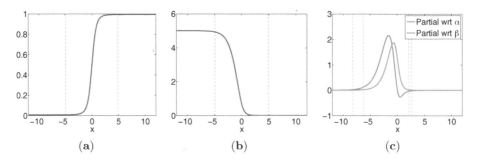

(a)   (b)   (c)

Figure 1.9: Classification of points on the x-axis, with (a) the decision region: $\hat{p}(x) = \ell_5(\sigma(\alpha x - \beta))$ with $\alpha = 1$, $\beta = 0$; (b) the cross entropy: $f(x) = -\log \hat{p}(x)$; (c) partial derivatives of the loss function with respect to α and β.

small. For points in the transition region the change will be relatively large, and the gradient at those points will therefore be larger. This suggests that training points away from the transition region provide little information when deciding in which direction to move the hyperplane and how sharp the transition should be; this information predominantly comes from the training points in the transition region.

More formally, consider the minimization of the negative log likelihood loss function for this network, given by

$$f(x) = -\log \hat{p}(x) \qquad \text{with} \qquad \hat{p}(x) = \ell_5(\sigma(\alpha x - \beta)).$$

For the gradient, we need the derivative of the sigmoid function, $\sigma'_\gamma(x) = 2\ell'_\gamma(x)$ with

$$\ell'_\gamma(x) = \frac{\gamma e^{-\gamma x}}{(1 + e^{-\gamma x})^2} = \gamma \left(\frac{1 + e^{-\gamma x}}{(1 + e^{-\gamma x})^2} - \frac{1}{(1 + e^{-\gamma x})^2} \right) = \gamma[\ell_\gamma(x) - \ell^2_\gamma(x)],$$

and the derivative of the negative log of the logistic function:

$$\frac{d}{dx}[-\log(\ell_\gamma(x))] = -\frac{\ell'_\gamma(x)}{\ell_\gamma(f(x))} = -\gamma \frac{\ell_\gamma(x) - \ell^2_\gamma(x)}{\ell_\gamma(x)} = \gamma[\ell_\gamma(x) - 1].$$

Combining the above we have

$$\partial f / \partial \alpha = \gamma[\ell_\gamma(\sigma(\alpha x - \beta)) - 1] \cdot \sigma'(\alpha x - \beta) \cdot x$$
$$\partial f / \partial \beta = -\gamma[\ell_\gamma(\sigma(\alpha x - \beta)) - 1] \cdot \sigma'(\alpha x - \beta),$$

with $\gamma = 5$. The loss function and partial derivatives with respect to α and β are plotted in Figure 1.9(b) and (c). The vertical lines in plot (c) indicate where the gradients fall below one percent of their asymptotic value. As expected, points beyond these lines do indeed contribute very little to the gradient, regardless of whether they are on the right or the wrong side of the hyperplane.

1.4.2 General mechanism

For the contribution of each sample to the gradient in general settings we need to take a detailed look at the backpropagation process. This is best illustrated using a concrete three-layer neural network:

$$A_1 = \begin{bmatrix} 1.0 & 0.3 \\ 0.4 & -1.0 \end{bmatrix} \qquad b_1 = \begin{bmatrix} -1.0 \\ 0.5 \end{bmatrix} \qquad \nu_1 = \sigma_3,$$

$$A_2 = \begin{bmatrix} -1 & -1 \\ 1 & -1 \\ 1 & 1 \\ -1 & 1 \end{bmatrix} \qquad b_2 = \begin{bmatrix} 1 \\ 1 \\ 1 \\ 1 \end{bmatrix} \qquad \nu_2 = \sigma_3, \tag{1.8}$$

$$A_3 = \begin{bmatrix} 1 & 0 & 1 & 0 \\ 0 & 1 & 0 & 1 \end{bmatrix} \qquad b_3 = \begin{bmatrix} -1.1 \\ -1.1 \end{bmatrix} \qquad \nu_3 = \mu.$$

Denoting by $f(x)$ the negative log likelihood of $\mu(x)$, the forward and backward passes through the network can be written as

$$
\begin{aligned}
v_1 &= A_1 x_0 - b_1 \\
x_1 &= \sigma_3(v_1) & y_1 &= \frac{\partial x_1}{\partial v_1} \cdot z_2 = \sigma_3'(v_1) \cdot z_2 \\
v_2 &= A_2 x_1 - b_2 & z_2 &= \frac{\partial v_2}{\partial x_1} \cdot y_2 = A_2^T y_2 \\
x_2 &= \sigma_3(v_2) & y_2 &= \frac{\partial x_2}{\partial v_2} \cdot z_3 = \sigma_3'(v_2) \cdot z_3 \\
v_3 &= A_3 x_2 - b_3 & z_3 &= \frac{\partial v_3}{\partial x_2} \cdot y_3 = A_3^T y_3 \\
x_3 &= f(v_3) & y_3 &= \frac{\partial x_3}{\partial v_3} = \frac{\partial f}{\partial v_3} = \nabla f(v_3),
\end{aligned}
\tag{1.9}
$$

where the left and right columns respectively denote the stages in the forward and backward pass. The regions formed during the forward pass are shown in Figure 1.10. With this, the partial differentials with respect to weight

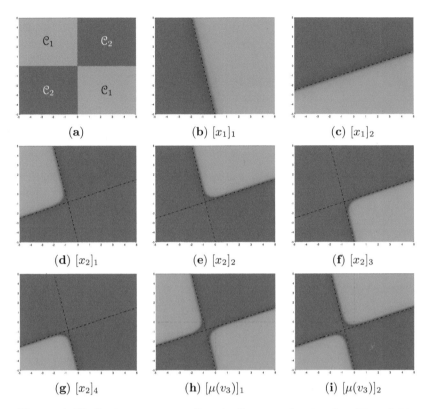

Figure 1.10: Regions corresponding to the neural network with weights defined by (1.8), with (a) the two ground-truth classes; (b,c) output regions of the first layer; (d–g) output regions of the second layer; and (h,i) output regions of the third layer (showing only the intermediate output $\mu(v_3)$ instead of the loss function output) with desired class boundaries superimposed.

matrices and bias vectors are of the following form:

$$\frac{\partial f}{\partial [A_3]_{i,j}} = \frac{\partial v_3}{\partial [A_3]_{i,j}} \cdot y_3 = [x_2]_j \cdot [y_3]_i, \quad \text{and}$$

$$\frac{\partial f}{\partial b_3} = \frac{\partial v_3}{\partial b_3} \cdot \frac{\partial f}{\partial v_3} = \frac{\partial v_3}{\partial b_3} \cdot y_3 = -y_3.$$

(1.10)

We now analyze each of the backpropagation steps to explain the relationship between the regions of high and low confidence at each of the neural network layers and the gradient values or importance of different points in the feature space. In all plots we only show the absolute values of the quantities of interest because we are mostly interested in the relative magnitudes over the feature space rather than their signs.

Backpropagation through the third layer   After a forward pass through the network we can evaluate the loss function and its gradients, shown in Figure 1.11(a). In this particular example we have $[y_3]_1 = -[y_3]_2$, so we only show the former. Given y_3 we can use (1.10) to compute the partial differentials of f with respect to the entries in A_3 and b_3. The partial differential with respect to b_3 simply coincides with $-y_3$, and is therefore not very interesting. On the other hand, we see that the partial differential with respect to $[A_3]_{i,j}$ is formed by multiplying $[y_3]_i$ with the output value $[x_2]_i$. When looking at the feature space representation for the specific case of $[A_3]_{1,2}$ and using absolute values, this corresponds to the pointwise multiplication of the values in Figure 1.11(a) with the mask shown Figure 1.11(b). This multiplication causes the partial differential to be reduced in areas of low confidence in $[x_2]_2$. In addition, it causes the partial differential to vanish at points at the zero crossing of the boundary regions, as illustrated by the white curve in the upper-right corner of Figure 1.11(c). Finally, we multiply y_3 by the transpose of A_3 to obtain z_3, the gradient with respect to the input of the third layer, shown in Figure 1.11(d).

Backpropagation through the second layer   Given z_3 we can backpropagate through the nonlinearity of the second layer by multiplication with $\sigma_3'(v_2)$ to obtain y_2. When considering the backpropagation of a single neuron with points from the entire feature space, this amounts to multiplication by a mask, as shown in Figure 1.11(e). The mask is formed by computing the gradient of the sigmoid at v_2. As illustrated in Figure 1.1(a), the gradient of the sigmoid is a kernel around the origin. When applied to v_2 (the preimage of x_2 under σ_3) it emphasizes the regions of low confidence and suppresses the regions of high confidence, as can be seen when comparing the mask for $[v_2]_2$, shown in Figure 1.11(e), with the corresponding region $[x_2]_2$ shown in Figure 1.10(e). The result obtained with multiplication by the mask is illustrated in Figure 1.11(f) and shows that backpropagation of the error is most predominant in the boundary region as well as in some regions where it was large to start with. Because of the shape of the mask we also see that the gradient $[y_2]_2$ nearly vanishes for points in the bottom-left quadrant.

From y_2 we can compute the partial differential with respect to the entries of A_2 through multiplication by x_1, which again damps values around the transition region, and backpropagate further to get z_2, as shown in Figures 1.11(g)–(i).

Backpropagation in the first layer   To obtain y_1, we need to multiply z_2 by the mask corresponding to the preimage of the regions in x_1. Unlike all other layers, these values are unbounded in the direction of the hyperplane normal

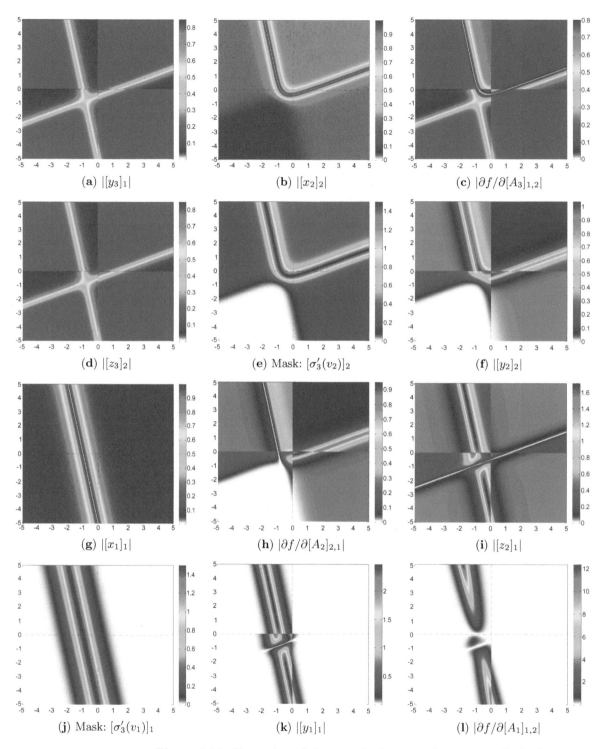

Figure 1.11: Illustration of the error backpropagation process. Each plot shows how the given quantity varies for points over the entire feature space.

and, as shown in Figure 1.11(j), result in masks that vanish away from the boundary region. Multiplication by the mask corresponding to $[v_1]_1$ gives $[y_1]_1$ shown in Figure 1.11(k). We finally obtain the partial differentials with respect to the entries in A_1 by multiplying by the corresponding entries in x_0. For the first layer this stage actually amplifies the gradient entries whenever the corresponding coordinate value exceeds one in absolute value. In subsequent layers the values of x lie in the -1 to 1 output range of the sigmoid function and can therefore only reduce the gradient components.

Gradients and scaling of the weights In Figure 1.12(a) we plot the maximum absolute gradient components for each of the three weight matrices. It is clear that the partial differentials with respect to A_3 are predominant in misclassified regions, but also exist outside of this region in areas where the objective function could be minimized further by increasing the confidence levels (scaling up the weight and bias terms). In the second layer, the backpropagated values are damped in the regions of high confidence and concentrate around the decision boundaries, which, in turn, are aligned with the underlying hyperplanes. Finally, in the first layer, we see that gradient values away from the hyperplanes have mostly vanished as a result of multiplication with the sigmoid gradient mask, despite the multiplication with potentially large coordinate values. Overall we see the tendency of the gradients to become increasingly localized in feature space towards the first layer. The boundary shifts we discussed in Section 1.2.4 can lead to additional damping, as the sigmoid derivative masks no longer align with the peaks in the gradient field. Scaling of the weight and bias is detrimental to the backpropagation of the error (a phenomenon that is also known as saturation of the sigmoids (15)) and can lead to highly localized gradient values. This is illustrated in Figures 1.12(b) and (c) where we scale all weight and bias terms by a factor of 2 and 3, respectively. Especially in deep networks it can be seen that a single sharp mask in one of the layers can localize the backpropagating error and thereby affect all preceding layers. These figures also show that the increased scaling of the weights not only leads to localization, but also to attenuation of the gradients. In the first layer this is further aided by the multiplication with the coordinate values. Summarizing, we see that the repeated multiplication by the masks generated by the derivative of the activation function tends to localize gradients. Multiplication with A^T in the back propagation mixes the regions with large gradients, but the location of these regions does not otherwise change. Finally, we note that the above principles are not restricted to the sigmoid or hyperbolic tangent functions. However, for activation functions where the gradient masks are not localized, for example for rectified linear units of the form $\max(\alpha x, \beta x)$, the vanishing gradient problem is less of a problem.

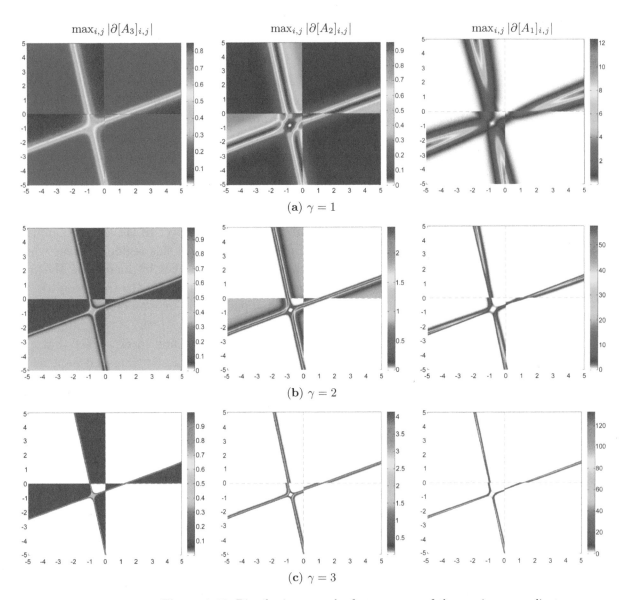

Figure 1.12: Distribution over the feature space of the maximum gradient components for the weight matrices in each of the three layers. The weight and bias terms in all layers are scaled by a factor γ.

1.5   Optimization

In the previous section we studied how individual training samples contribute to the overall gradient (1.7) of the loss function. In this section we take a closer look at the dynamic behavior and the changing relevance of training samples during optimization over the network parameter vector s.

The parameter updates are gradient-descent steps of the form

$$s^{k+1} = s^k - \alpha \nabla \phi(s^k), \tag{1.11}$$

with learning rate α. The goal of this section is to clarify the relationships between the training set and the optimization process of the network parameters. To keep things simple we make no effort to improve the efficiency of the optimization process and, unless noted otherwise, we use a fixed learning rate with a moderate value of $\alpha = 0.01$. Likewise, we compute the exact gradient using the entire training set instead of using an approximation based on suitably chosen subsets, as is done in practically favored stochastic gradient descent (SGD) methods[3]. Throughout this section we place a particular emphasis on the first layer of the neural network. To illustrate certain mechanisms it often helps to keep parameters of subsequent layers fixed. In this case it is implied that the corresponding entries in the gradient update in (1.11) are zeroed out.

1.5.1 Sampling density and transition width

To investigate the roles of sampling density and transition width we start with a very simple example with feature vectors $x \in [-6, 6] \times [-1, 1] \subset \mathbb{R}^2$, and two classes: one to the left of the y-axis (first entry is negative), and one to the right. Example training sets with samples in each of the two classes are plotted in Figure 1.13(a) and (b). Given such training sets we want to learn the classification using a neural network with a single hidden layer consisting of one node. To ensure that the classes are well defined we place four training samples—two for each class– near the interface of the two classes and sample the remaining points to the left and right of these points. Unless stated otherwise we keep all network parameters fixed except for the weights and bias terms in the first layer.

1.5.1.1 Sampling density

In the first experiment we study how the number or density of training points affects the optimization. We initialize the network with parameters

$$A_1 = \tfrac{25}{\sqrt{1.09}}[1, 0.3], \ b_1 = A_1 \cdot [2, 0]^T, \quad A_2 = [3; -3], \ b_2 = [0; 0],$$

---

3. The mechanisms exposed in this section are general enough to carry over to stochastic gradient descent and other methods without substantial changes.

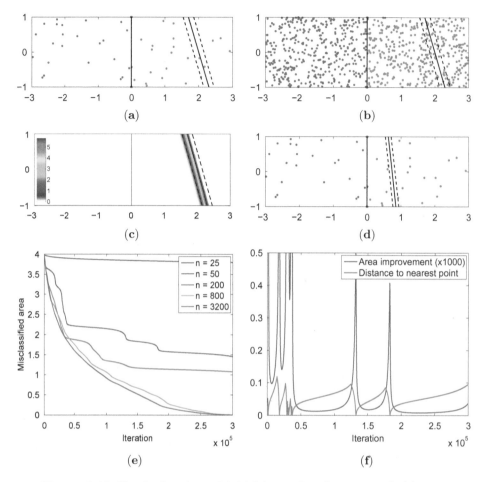

Figure 1.13: Simple domain and initial hyperplane location with (a) 50, and (b) 800 training samples equally divided over two classes, along with (c) the initial gradient field. Plot (d) shows the location of the hyperplane after 300,000 iterations and (e) shows the area of the misclassified region as a function of iteration for different numbers of training samples. Plot (f) shows the reduction in misclassified area per iteration and the distance between the hyperplane and the nearest sample when 50 training samples are used.

and keep the parameters in the second layer fixed. Parameter b_1 is chosen such that the initial hyperplane goes through the point $(2, 0)$. Training sets consist of n samples, including the four at the interface, and are chosen such that the number of points in each class differs by at most one. Figures 1.13(a) and (b) illustrate such sets for $n = 50$, and $n = 800$, respectively. The hyperplane is indicated by a thick black line, bordered with two dashes lines which indicate the location where the output of the first layer is equal to ± 0.95. Figure 1.13(c) shows the magnitude of the partial differential with respect to $[A_1]_1$ at the initial parameter setting over the entire domain.

The gradient $\nabla\phi(s)$ is then computed as the average of the gradient values evaluated at the individual training samples. As a measure of progress we can look at the area of the misclassified region, i.e., the region between the y-axis and the hyperplane (note this quantity does not include information about the confidence levels of the classification). Figure 1.13(e) shows this area as a function of iteration for different sampling densities. The shape of the loss function curves are very similar to these and we therefore omit them here. For $n = 800$ and $n = 3200$ the area of the misclassified region steadily goes down to zero, although the rate at which it does so gradually diminishes. Although not apparent from the curves, this phenomenon happens for all the training sets used here and we will explain exactly why this happens in Section 1.5.2. Progress for $n = 50$ and $n = 200$ appears much less uniform and exhibits pronounced stages of fast and slow progress. The reason for this is a combination of the sampling density and the localized gradient. From Figure 1.13(c) we can see that the gradient field is concentrated around the hyperplane, with peak values slightly to the left of the hyperplane. When the sampling density is low it may happen that none of the training samples is close to the hyperplane. When this happens, the gradient will be small, and consequently progress will be slow. When one or more points are close to the hyperplane, the gradient will be larger and progress is faster. Figure 1.13(f) shows the rate of change in the area of the misclassified region along with the distance between the hyperplane and its nearest training sample for $n = 50$. It can be seen that the rate increases as the hyperplane moves towards the training sample, with the peak rate happening just before the hyperplane reaches the point. After that the rate gradually drops again as the hyperplane slowly moves further away from the sample. This is precisely the state at 300,000 iterations, which is illustrated in Figure 1.13(d). For $n = 25$, we find ourselves in the same situation right at the start. Initially we move away from a single training point, but as a consequence of the low sampling density, no other sampling points are nearby, causing a prolonged period of very slow progress. The discrete nature of training samples is less pronounced when the overall sampling density is high, or when the transition widths are large.

1.5.1.2   *Transition width*

To illustrate the effect of transition widths, we used the setting with 3,200 samples as described above, but scaled the row vector of the initial A_1 to have Euclidean norm ranging from 1 to 100. In each case we adjust b_1 such that the initial hyperplane goes through the point $(2, 0)$. As shown in Figure 1.14(a), the misclassified area reaches zero almost immediately when A_1 is scaled to have unit norm. In other words, the hyperplane is placed

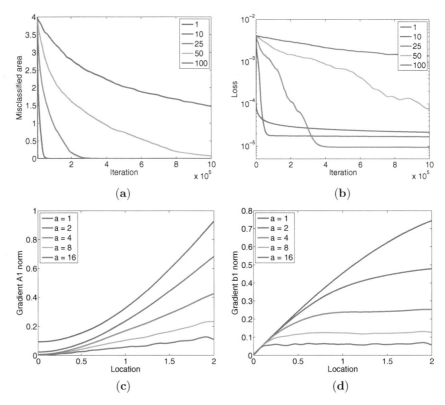

Figure 1.14: Plots of (a) misclassified area and (b) the value of the loss function as a function of iterations with the norm of weights A_1 ranging from 1 to 100. Norm of the gradients with respect to (c) A_1 and (d) b_1 as a function of hyperplane location with $A_1 = [a, 0]$ for different values of a.

correctly in this case after only 3,260 iterations. As the norm of the initial A_1 increases, it takes longer to reach this point: for an initial norm of 10 it takes some 72,580 iterations, whereas for an initial norm of 25 it takes over 300,000. Accordingly, we see from Figure 1.14(b) that the loss also drops much faster for small weights than it does for large weights. However, once the hyperplane is in place, the only way to decrease the loss is by scaling the weights to improve the confidence. This process can be somewhat slow when the weights are small and the hyperplane placement is finalized (as is the case when we start with small initial weights). As a result, the setup with initial weight of 25 eventually catches up with the earlier two, simply because it has a much sharper transition at the boundary as the hyperplane finally closes in to the right location.

The reason why the hyperplane moves faster for small initial weights is twofold. First, the transition width and support of the gradient field are larger. As a result, more sample points contribute to the gradient, leading to a larger overall gradient value. This is shown in Figures 1.14(c) and (d) in which we plot the norm of the gradients with respect to A_1 and b_1 when choosing $A_1 = [a, 0]$, and b such that the hyperplane goes through the given location on the x-axis. The gradients with respect to either parameters are larger for smaller a. (Unlike in Figure 1.12, the localization of the gradient here is due only to scaling of the weights in the first layer; the intensity of the gradient field therefore remains unaffected.) As the value of a increases, the curves in Figures 1.14(c) become more linear. For those values the gradient is highly localized and, aside from the scaling by the training point coordinates, largely independent of the hyperplane location. The gradient with respect to b_1 does not include this scaling and therefore remains nearly constant as long as the overlap between the transition width and the class boundary is negligible. As the hyperplane moves into the right place, the gradient vanishes due to the cancellation of the contributions from the training points from the classes on either side of it. The curves for $a = 16$ and, to a lesser extent for $a = 8$, show minor aberrations due to a relatively low sampling density compared to the transition width. Second, having larger gradient values for smaller weights means that the relative changes in weights are amplified, thereby allowing the hyperplane to move faster.

1.5.2 Controlling the parameter scale

In this section we work with a modified version of the domain shown in Figure 1.13(a). In particular, we change the horizontal extent from $[-3, 3]$ to $[-30, 30]$, and randomly select 250 training samples uniformly at random for each of the two classes (thus leaving the sampling density unaffected compared to the original $n = 50$). As a first experiment we optimize a three-layer network with initial parameters:

$$
\begin{aligned}
A_1 &= [1, 0.3]/\sqrt{1.09}, & b_1 &= A_1 \cdot [25; 0], \\
A_2 &= 3, & b_2 &= 0, \\
A_3 &= [3; -3], & b_3 &= 0.
\end{aligned}
\tag{1.12}
$$

When we look at the row-norms of the weight matrices, plotted in Figure 1.15(a), we can see that all of them are growing. This growth can help improve the final confidence levels, but can be detrimental during the optimization process, especially when it occurs in the layers between the first and the last. Indeed, we can see from Figure 1.15(b) that the hyperplane never quite reaches the origin, despite the large number of iterations. As illustrated

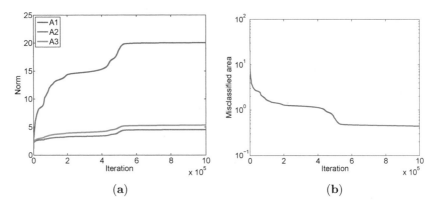

Figure 1.15: Plots of (a) growth in the norms of the weight matrices and (b) reduction of the misclassified area as a function of iteration.

in Figure 1.12, scaling of the weight and bias terms leads to increasingly localized gradients. When the training sample density is low compared to the size of the regions where the gradient values are significant, it can easily happen that no significant values from the gradient field are sampled into the gradient. This applies in particular to the first several layers (depending on the network depth) where the gradient fields become increasingly localized (though not necessarily small) as a result of the sigmoidal gradient masks that are applied during back propagation, along with shifts in the boundary regions. This 'vanishing gradient' phenomenon can prematurely bring the training process to a halt; not because a local minimum is reached, but simply because the sampled gradient values are excessively small[4]. Scaling of the parameters in any layer except the last can cause the gradient field to become highly localized for the current and all preceding layers. This can cause a cascading effect in which suboptimal parameters in a stalled first layer lead to further parameter scaling in later layers, eventually causing the second layer to stall, and so on. To avoid this, we need to control the parameter scale during optimization.

Parameter growth can be controlled by adding a regularization or penalty term to the loss function, or by imposing explicit constraints. Extending (1.4) we could use

$$\underset{s}{\text{minimize}} \quad \phi(s) + r(s), \quad \text{or} \quad \underset{s}{\text{minimize}} \quad \phi(s) \\ \text{subject to} \quad c_i(s) \leq 0, \tag{1.13}$$

---

4. Small gradients can also be due to cancellations in the various contributions. In practice, and especially when classes mix in a boundary zone, the small gradient can be expected to be due to a combination of two effects.

where $r(s)$ is a regularization function, and $c_i(s)$ are constraint functions. The discussions so far suggest some natural choices of functions for different layers. The function in the first layer should generally be based on the (Euclidean) ℓ_2 norm of each of the rows in A_1, such as their sum, maximum, or ℓ_2 norm. The reason for this is that each row in A_1 defines the normal of a hyperplane, and using any function other than an ℓ_2 norm may introduce a bias in the hyperplane directions due to a lack of rotational invariance. For subsequent layers k (except possibly the last layer) we may want to ensure that the output cannot be too large. In the worst case, each input from the previous layer is close to $+1$ or -1, and we can limit the output value by ensuring that the sum of absolute values, i.e., the ℓ_1 norm, of each row in A_k is sufficiently small. Of course, the corresponding value in b_k could still be large, which may suggest adding a constraint that $\|[A_k]_j\|_1 \leq |[b_k]_j|$ for each row j. However, this constraint is non-convex and may impede sign changes in b. The use of an ℓ_1 norm-based penalty or constraint on intermediate layers has the additional benefit that it leads to sparse weight matrices, which can help reduce model complexity as well as evaluation cost.

As an illustration of the effect of ℓ_2 regularization on the first layer we consider the setting as given in (1.12), but with the second layer removed. We optimize the weight and bias terms in the first layer using the standard formulation (1.4), as well as those in (1.13) with $r(s) = \lambda/2\|A_1^T\|_2^2$ or $c(s) = \|A_1^T\|_2 \leq \kappa$. For simplicity we keep all other network parameters fixed. Optimization in the constrained setting is done using a basic gradient projection method with step size fixed to 0.01, as before. The results are shown in Figure 1.16. When using the standard formulation we see from Figure 1.16(a) that, like above and in Figures 1.13(a,d), the ℓ_2 norm of the row in A_1 keeps growing. This is explained as follows: suppose the hyperplane is vertical with A_1 of the form $[a, 0]$, and $b_1 = b$. Then the area of the misclassified region is $2|b|/|a|$. We can therefore reduce the misclassified area (and in this case the loss function) by increasing a and decreasing b, which is exactly what happens. However, from Figure 1.16(b) we can see that the rate at which the misclassified area is reduced decreases. The reason for this is a combination of three factors. First, the speed at which $|b|/|a|$ goes towards zero slows down as a gets larger. Second, the peak of the gradient field lies along the hyperplane and shifts towards the origin with it. Because the gradient in the first layer is formed by a multiplication of the backpropagated error with the feature vectors (coordinates), the gradient gets smaller too. Third, because of the growing norm of A_1, the transition width shrinks and causes the gradient to become more localized. As a result, fewer training points sample the gradient field at significant values, leading to smaller overall gradients with respect to both A_1 and b_1.

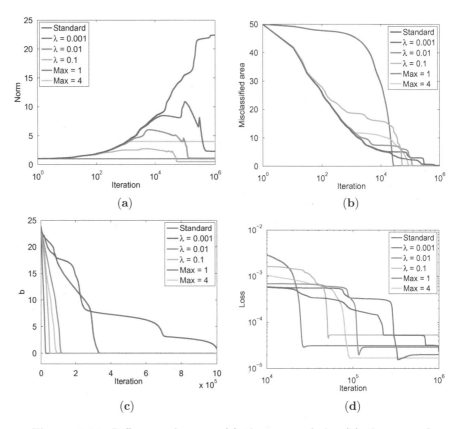

Figure 1.16: Differences between (a) the norm of A_1; (b) the area of the misclassified region; (c) the magnitude of b; and (d) the loss function, as a function of iteration for standard gradient descent and variations of regularized and constrained optimization.

There is not much we can be do about the first two causes, but adding a regularization term or imposing constraints, certainly does help with the third, and we can see from Figure 1.16(a) that the norm of A_1 indeed does not grow as much as in the standard approach. At first glance, this seems to hamper the reduction of the misclassified area, shown in Figure 1.16(b). This is true initially when most of the progress is due to the scaling of A_1, however, the moderate growth in A_1 also prevents strong localization of the gradient and therefore results in much steadier reduction of b, as shown in Figure 1.16(c). The overall effect is that the constrained and regularized methods catch up with the standard method and reduce the misclassified area to zero first. Even so, when looking at the values of the loss function without the penalty term, as plotted in Figure 1.16(d), we see that the standard method still reaches the lowest loss value, even though all methods have zero misclassification. As before, this is because the two classes are disjoint and are best separated with a very sharp transition. The order in which the lines in Figure 1.16(d) appear at the end, is therefore related to the norms in Figure 1.16(a). This suggests the use of cooling or continuation strategies in which norms are gradually allowed to increase. The initial small weights ensure that many of the training samples are informative and contribute to the gradients of all layers, thereby allowing the network to find a coarse class alignment. From there the weights can be allowed to increase slowly to fine tune the classification and increase confidence levels. Of course, while doing so, care needs to be taken not to allow excessive scaling of the weights, as this can lead to overfitting.

Instead of scaling weight and bias terms we could also consider scaling sigmoid parameters γ, or learn them (22). One interesting observation here is that even though all networks with parameters αA_1, αb_1, and γ/α are equivalent for $\alpha > 0$, their training certainly is not. The reason is the $1/\alpha$ term that applies to the gradients with respect to A_1 and b_1. Choosing $\alpha > 1$ means larger parameter values and smaller gradients. This reduces both the absolute and relative change in parameter values and is equivalent to having a stepsize that is α^2 smaller. Instead of doing joint optimization over both the layer and nonlinearity parameters, it is also possible to learn the nonlinearity parameters as a separate stage after optimization of the weight and bias terms.

1.5.3   Subsampling and partial backpropagation

Consider the scenario shown in Figure 1.13(a) and suppose we double the number of training samples by adding additional points to the left and right of the current domain. In the original setting, the gradient with respect to the weights in the first layer is obtained by sampling the gradient field shown in Figure 1.13(c). In the updated setting, all newly added points are located away from the decision boundary. As a result, their contribution to the gradient is relatively small and the overall gradient may be very similar to the original setting. However, because the loss function $\phi(s)$ in (1.4) is defined as the average of the individual loss-function components, we now need to divide by $2N$ rather than N, thereby effectively scaling down the gradient by a factor of approximately two. Another way to say this is that the stepsize is almost halved by adding the new points. This example is of course somewhat contrived, since additional training samples can typically be expected to follow the same distribution as existing points and therefore increase sampling density. Nevertheless, this example makes one wonder whether the training samples on the left and right-most side of the original domain are really needed; after all, using only the most informative samples in the gradient essentially amounts to larger stepsize and possibly a reduction in computation.

For sufficiently deep networks with even moderate weights, the hyperplane learning is already rather myopic in the sense that only the training points close enough to the hyperplane provide information on where to move it. This suggests a scheme in which we subsample the training set and for one or more iterations work with only those points that are relevant. We could for example evaluate $v_1 = A_1 x_0 - b_1$ for each input sample x_0, and proceed with the forward and backward pass only if the minimum absolute entry in v_1 is sufficiently small (i.e., the point lies close enough to at least one of the hyperplanes). This approach works to some extend for the first layer when the remaining layers are kept fixed, however, it does not generalize because the informative gradient regions can differ substantially between layers (see e.g., Figure 1.12). Instead of forming a single subsampled set of training points for all layers we can also form a series of sets—one for each layer—such that all points in a set contribute significantly to the gradient for the corresponding and subsequent layers. This allows us to appropriately scale the gradients for each layer. It also facilitates partial backpropagation in which the error is backpropagated only up to the relevant layer, thereby reducing the number of matrix-vector products. Given a batch of points, we could determine the appropriate set by evaluating the gradient contribution to each layer and finding the lowest layer for which the contribution is above some threshold. Alternatively, we could use the following partial backpropagation approach, which may be beneficial in its own right, especially for deep networks.

In order to do partial backpropagation, we need to determine at which layer to stop. If this information is not given a priori, we need a conservative and efficient mechanism that determines if further backpropagation is warranted. One such method is to determine an upper bound on the gradient components of all layers up to the current layer and decide if this is sufficiently small. We now derive bounds on $\|\partial f / \partial A_k\|_F$ and $\|\partial f / \partial b_k\|_F$ as well as on $\max_{i,j} |[\partial f / \partial A_k]_{i,j}|$ and $\|\partial f / \partial b_k\|_\infty$. It easily follows from (1.10) that these quantities are equal to $\|x_{k-1}\|_2 \|y_k\|_2$ and $\|y_k\|_2$, respectively $\|x_{k-1}\|_\infty \|y_k\|_\infty$ and $\|y_k\|_\infty$. Since x_{k-1} is known explicitly from the forward pass, it suffices to bound the norms of y_k. In fact, what we are really after is to bound the norms of y_k for all $1 \leq k < j$ given y_j, since we can stop backpropagation only if all of them are sufficiently small. For the ℓ_2 norm we have

$$
\begin{aligned}
\|y_{k-1}\|_2 \;&\leq\; \|\sigma'_{\gamma_{k-1}}(v_{k-1})\|_\infty \|z_k\|_2 \\
&\leq\; \sigma'_{\gamma_{k-1}}([v_{k-1}]_i) \cdot \sigma_{\max}(A_k) \|y_k\|_2,
\end{aligned} \tag{1.14}
$$

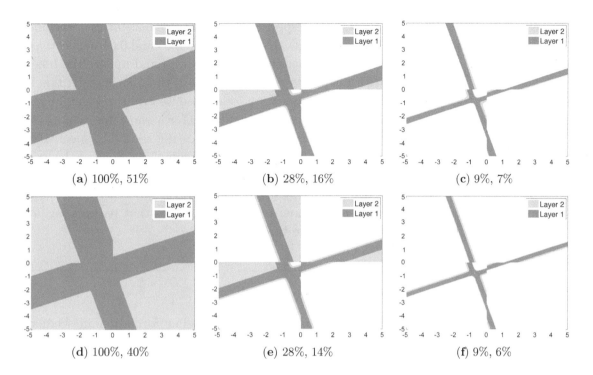

Figure 1.17: Regions of the feature space that are backpropagated to layers 2 and 1. From left to right we have the settings $\gamma = 1$, $\gamma = 2$, and $\gamma = 3$ from Figure 1.12, respectively. The top row shows the results obtained with the Frobenius norm of the gradients with respect to the weight matrices in each layer. The bottom row shows the results obtained by bounding the gradients elementwise. The percentages indicate the fraction of the feature space that was backpropagated to the second, and first layer.

where $i := \arg\min_j |[v_{k-1}]_j|$, and $\sigma_{\max}(A_k)$ is the largest singular values of A_k. Once we have a bound on $\|y_j\|_2$ we can apply (1.14) with $k = j$ to bound $\|y_{j-1}\|_2$. Although computation of $\sigma_{\max}(A_k)$ needs to be done only once per batch but may still be prohibitively expensive. In practice, however, it may suffice to work with an approximate value, or use an alternative bound instead. For ℓ_∞ we find

$$
\begin{aligned}
\|y_{k-1}\|_\infty &\leq \max_i \{\sigma'_{\gamma_{k-1}}([v_{k-1}]_i) \cdot \|[A_k]_i\|_2 \|y_k\|_2\} \\
&\leq \|\sigma'_{\gamma_{k-1}}(v_k)\|_\infty \|y_k\|_2 \max_i\{\|[A_k]_i\|_2\},
\end{aligned}
\tag{1.15}
$$

where the second, looser bound can be used if we want to avoid evaluating $\sigma'_{\gamma_{k-1}}$ for all entries in v_k; the infinity norm of this vector can be evaluated as above.

We applied the second bound in (1.14) and the first bound in (1.15) to the setting for Figure 1.12 as follows. We first compute y_3 and evaluate the bound the gradients with respect to the weight and bias terms in the first and second layer. If these bounds are smaller than 0.05 and 0.01, respectively, we stop backpropagation. Otherwise, we evaluate y_2 and update the bound on the gradient with respect to the parameters of the first layer. If this is less than 0.05 we stop backpropagation, otherwise we evaluate y_1 and complete the backpropagation process. In Figure 1.17 we show the regions of the feature space where backpropagation reaches the second, respectively first layer. These regions closely match the predominant regions of the gradient fields shown in Figure 1.12. In practical applications the threshold values could be based on previously computed (partial) gradient values, and may be adjusted when the number of training samples that backpropagate to a given layer falls below some threshold.

1.6 Conclusions

We reviewed and studied the decision region formation in feedforward neural networks with sigmoidal nonlinearities. Although the definition of hyperplanes and their subsequent combination is well known, very little attention has so far been given to transitions regions at the boundaries of classes and other regions with varying levels of classification confidence. We clarified the relation between the scaling of the weight matrices, the increase in confidence and sharpening of the transition regions, and the corresponding localization of the gradient field. The degree of localization differs per layer and is one of the main factors that determine how much progress can be made at each step of the training process: a high level of localization combined with a relatively coarse sampling density or small batch size leads to the vanishing gradient problem where updates to one or more layers become excessively small. The gradient field tends to become increasingly localized towards the first layer, and the parameters in this layer are therefore most likely to get stuck prematurely. When this happens, subsequent layers must form classifications regions based on suboptimal hyperplane locations. It is often possible to slightly decrease the loss function by increasing confidence levels by scaling parameters in later layers. This can lead to a cascading effect in which layers successively get stuck. The use of regularized or constrained optimization can help control the scaling of the weights, thereby limiting the amount of gradient localization and thus avoiding or reducing these problems. By gradually allowing the weights to increase it is possible to balance progress in the learning process and attaining decision regions with sufficiently high confidence levels. In addition, regularized and constrained optimization can help prevent overfitting. Analysis of the gradient field also shows that at any given iteration, the contributions of different training points to the gradient can vary substantially. Localization of the gradient towards the first layer also means that some points are informative only from a certain layer onwards. Together this suggests dynamic subset selection and partial backpropagation, or adaptive selection of the step size for each layer depending on the number of relevant points.

We hope that some of the results presented in this work will contribute to a better understanding of neural networks and eventually lead to new or improved algorithms. There remain several topics that are interesting but beyond the scope of the present chapter. For example, it would be interesting to see what the hyperplanes generated during pre-training using restricted Boltzmann machines (11) look like, and if there are better choices. One possible option is to select random training samples from each class and generate randomly oriented hyperplanes through these points by appropriate choice of b. Likewise, given a hyperplane orientation and a desired class, it is also possible to place the hyperplane at the class boundary by choosing b to coincide with the largest or smallest inner product of the normal with points from that class. Another interesting topic is an extension of this work to other nonlinearities such as the currently popular rectified linear unit given by $\nu(x) = \max(0, x)$. The advantage of these units is that gradient masks is one for all all positive inputs and are not localized, thereby avoiding gradient localization and thus allowing the error to backpropagate more easily. It would be interesting to look at the mechanisms involved in the formation of decision regions, which differ from those of sigmoidal units. For example, it is not entirely clear how the logical AND should be implemented: summing inverted regions and thresholding may work in some cases, but more generally it should consist of the minimum of all input regions. In terms of combinatorial properties, bounds on the number of regions generated using neural networks with rectified and piecewise linear functions were recently obtained in (17, 19). The main problem with rectified linear units is that it maps all negative inputs to zero, thereby creating a zero gradient mask at those locations. The softplus nonlinearity (9), which is a smooth alternative in which the gradient mask never vanishes, would also be of interest. Finally it would be good to get a better understanding of dropout (12) and second-order methods from a feature-space perspective.

References

1. M. Anthony. Boolean functions and artificial neural networks. Technical Report CDAM research report series, LSE-CDAM-2003-01, Centre for Discrete and Applicable Mathematics, London School of Economics and Political Science, London, UK, 2003.

2. C. M. Bishop. *Neural Networks for Pattern Recognition.* Oxford University Press, Inc., New York, NY, USA, 1995.

3. R. P. Boland and J. Urrutia. Separating collections of points in Euclidean spaces. *Information Processing Letters*, 53(4):177–183, February 1995.

4. E. M. Bronshteyn and L. D. Ivanov. The approximation of convex sets by polyhedra. *Sibirian Mathematical Journal*, 16(5):852–853, 1975.

5. E. M. Bronstein. Approximation of convex sets by polytopes. *Journal of Mathematical Sciences*, 153(6):727–762, 2008.

6. G. H. L. Cheang and A. R. Barron. A better approximation for balls. *Journal of Approximation Theory*, 104(2):183–203, 2000.

7. D. L. Donoho and M. Elad. Optimally sparse representation in general (nonorthogonal) dictionaries via ℓ^1 minimization. *Proceedings of the National Academy of Sciences*, 100(5):2197–2202, March 2003.

8. R. M. Dudley. Metric entropy of some classes of sets with differentiable boundaries. *Journal of Approximation Theory*, 10(3):227–236, 1974.

9. X. Glorot, A. Bordes, and Y. Bengio. Deep sparse rectifier networks. In *Proceedings of the 14th International Conference on Artificial Intelligence and Statistics*, volume 15, pages 315–323. JMLR W&CP, 2011.

10. G. Hinton, L. Deng, D. Yu, G. E. Dahl, A. Mohamed, N. Jaitly, A. Senior, V. Vanhoucke, P. Nguyen, T. N. Sainath, and B. Kingsbury. Deep neural networks for acoustic modeling in speech recognition. *IEEE Signal Processing Magazine*, 29(6):82–97, November 2012.

11. G. E. Hinton, S. Osindero, and Y.-W. Teh. A fast learning algorithm for deep belief nets. *Neural Computation*, 18:1527–1554, 2006.

12. G. E. Hinton, N. Srivastava, A. Krizhevsky, I. Sutskever, and R. R. Salakhutdinov. Improving neural networks by preventing co-adaptation of feature detectors. *The Computing Resarch Repository (CoRR)*, abs/1207.0580, 2012.

13. W. Y. Huang and R. P. Lippmann. Neural net and traditional classifiers. In D. Z. Anderson, editor, *Neural Information Processing Systems*, pages 387–396, 1988.

14. K. Böröczky Jr. and G. Wintsche. Covering the sphere by equal spherical balls. In B. Aronov, S. Bazú, M. Sharir, and J. Pach, editors, *Discrete and Computational Geometry – The Goldman-Pollak Festschrift*, pages 235–251. Springer, 2003.

15. Y. LeCun, L. Bottou, G. B. Orr, and K.-R. Müller. Efficient backprop. In G. B. Orr and K.-R. Müller, editors, *Neural Networks: Tricks of the Trade*, volume 1524 of *Lecture notes in computer science*, pages 9–50. Springer, 1998.

16. J. Makhoul, R. Schwartz, and A. El-Jaroudi. Classification capabilities of two-layer neural nets. In *14th International Conference on Acoustics, Speech and Signal Processing (ICASSP)*, pages 635–638, 1989.

17. G. Montúfar, R. Pascanu, K. Cho, and Y. Bengio. On the number of linear regions of deep neural networks. arXiv 1402.1869, February 2014.

18. J. Pach and G. Tardos. Separating convex sets by straight lines. *Discrete Mathematics*, 241(1–3):427–433, 2001.

19. R. Pascanu, G. Montúfar, and Y. Bengio. On the number of response regions of deep feed forward networks with piece-wise linear activations. arXiv 1312.6098, December 2013.

20. D. E. Rumelhart, G. E. Hinton, and R. J. Williams. Learning representations by back-propagating errors. *Nature*, 323:533–536, 1986.

21. L. Schläfli. *Theorie der Vielfachen Kontinuität*, volume 38 of *Neue Denkschriften der allgemeinen schweizerischen Gesellschaft für die gesamten Naturwissenschaften*. 1901.

22. A. Sperduti and A. Starita. Speed up learning and network optimization with extended back propagation. *Neural Networks*, 6(3):365–383, 1993.

23. H. Tverberg. A separation property of plane convex sets. *Mathematica Scandinavica*, 45:255–260, 1979.

24. A. D. Wyner. Random packings and coverings of the unit n-sphere. *Bell Labs Technical Journal*, 46(9):2111–2118, November 1967.

25. G. P. Zhang. Neural networks for classification: a survey. *IEEE Transactions on Systems, Man, and Cybernetics–Part C: Applications and Reviews*, 30(4):451–462, 2000.

2 Variable Selection in Gaussian Markov Random Fields

Jean Honorio   jhonorio@csail.mit.edu
CSAIL, MIT
Cambridge, MA, USA

Dimitris Samaras   samaras@cs.sunysb.edu
CS Department, Stony Brook University
Stony Brook, NY, USA

Irina Rish   rish@us.ibm.com
Computational Biology Center, IBM T. J. Watson Research Center
Yorktown Heights, NY, USA

Guillermo A. Cecchi   gcecchi@us.ibm.com
Computational Biology Center, IBM T. J. Watson Research Center
Yorktown Heights, NY, USA

We present a variable-selection structure learning approach for Gaussian graphical models. Unlike standard sparseness promoting techniques, our method aims at selecting the most-important variables besides simply sparsifying the set of edges. Through simulations, we show that our method outperforms the state-of-the-art in recovering the ground truth model. Our method also exhibits better generalization performance in a wide range of complex real-world datasets: brain fMRI, gene expression, NASDAQ stock prices and world weather. We also show that our resulting networks are more interpretable in the context of brain fMRI analysis, while retaining discriminability. From an optimization perspective, we show that a block coordinate descent method generates a sequence of positive definite solutions. Thus, we reduce the original problem into a sequence of strictly convex (ℓ_1, ℓ_p) regularized quadratic minimization subproblems for $p \in \{2, \infty\}$. Our algorithm is well founded since the optimal solution of the maximization problem is unique and bounded.

2.1   Introduction

Structure learning aims to discover the topology of a probabilistic graphical model such that this model represents accurately a given dataset. Accuracy of representation is measured by the likelihood that the model explains the observed data. From an algorithmic point of view, one challenge faced by structure learning is that the number of possible structures is super-exponential in the number of variables. From a statistical perspective, it is very important to find good regularization techniques in order to avoid overfitting and to achieve better generalization performance. Such regularization techniques will aim to reduce the complexity of the graphical model, which is measured by its number of parameters.

For Gaussian graphical models, the number of parameters, the number of edges in the structure and the number of non-zero elements in the inverse covariance or precision matrix are equivalent measures of complexity. Therefore, several techniques focus on enforcing sparseness of the precision matrix. An approximation method proposed in (25) relied on a sequence of sparse regressions. Maximum likelihood estimation with an ℓ_1-norm penalty for encouraging sparseness is proposed in (1, 8, 38).

In this chapter, summarizing our work presented earlier in (14), we assume a particular type of structured sparsity, where only a relatively small subset of network nodes is involved in "strong" interactions with some other nodes. Intuitively, we want to select these "important" nodes. However, the above methods for sparsifying network structure do not directly promote variable selection, i.e. group-wise elimination of all edges adjacent to an "unimportant" node. Variable selection in graphical models present several advantages. From a computational point of view, reducing the number of variables can significantly reduce the number of precision-matrix parameters. Moreover, group-wise edge elimination may serve as a more aggressive regularization, removing all "noisy" edges associated with nuisance variables at once, and potentially leading to better generalization performance, especially if, indeed, the underlying problem structure involves only a limited number of "important" variables. Finally, variable selection improves interpretability of the graphical model: for example, when learning a graphical model of brain area connectivity, variable selection may help to localize brain areas most relevant to particular mental states.

Our contribution is to develop variable-selection in the context of learning sparse Gaussian graphical models. To achieve this, we add an $\ell_{1,p}$-norm regularization term to the maximum likelihood estimation problem, for $p \in \{2, \infty\}$. We optimize this problem through a block coordinate descent

Table **2.1**: Notation used in this chapter.

Notation	Description
$\|\mathbf{c}\|_1$	ℓ_1-norm of $\mathbf{c} \in \mathbb{R}^N$, i.e. $\sum_n |c_n|$
$\|\mathbf{c}\|_\infty$	ℓ_∞-norm of $\mathbf{c} \in \mathbb{R}^N$, i.e. $\max_n |c_n|$
$\|\mathbf{c}\|_2$	Euclidean norm of $\mathbf{c} \in \mathbb{R}^N$, i.e. $\sqrt{\sum_n c_n^2}$
$\mathbf{A} \succeq \mathbf{0}$	$\mathbf{A} \in \mathbb{R}^{N \times N}$ is symmetric and positive semidefinite
$\mathbf{A} \succ \mathbf{0}$	$\mathbf{A} \in \mathbb{R}^{N \times N}$ is symmetric and positive definite
$\|\mathbf{A}\|_1$	ℓ_1-norm of $\mathbf{A} \in \mathbb{R}^{M \times N}$, i.e. $\sum_{mn} |a_{mn}|$
$\|\mathbf{A}\|_\infty$	ℓ_∞-norm of $\mathbf{A} \in \mathbb{R}^{M \times N}$, i.e. $\max_{mn} |a_{mn}|$
$\|\mathbf{A}\|_2$	spectral norm of $\mathbf{A} \in \mathbb{R}^{N \times N}$, i.e. the maximum eigenvalue of $\mathbf{A} \succ \mathbf{0}$
$\|\mathbf{A}\|_{\mathfrak{F}}$	Frobenius norm of $\mathbf{A} \in \mathbb{R}^{M \times N}$, i.e. $\sqrt{\sum_{mn} a_{mn}^2}$
$\langle \mathbf{A}, \mathbf{B} \rangle$	scalar product of $\mathbf{A}, \mathbf{B} \in \mathbb{R}^{M \times N}$, i.e. $\sum_{mn} a_{mn} b_{mn}$

method which yields sparse and positive definite estimates. We show that our method outperforms the state-of-the-art in recovering the ground truth model through synthetic experiments. We also show that our structures have higher test log-likelihood than competing methods, in a wide range of complex real-world datasets: brain fMRI, gene expression, NASDAQ stock prices and world weather. In particular, in the context of brain fMRI analysis, we show that our method produces more interpretable models that involve few brain areas, unlike standard sparseness promoting techniques which produce hard-to-interpret networks involving most of the brain. Moreover, our structures are as good as standard sparseness promoting techniques, when used for classification purposes.

Section 2.2 introduces Gaussian graphical models and techniques for learning them from data. Section 2.3 sets up the $\ell_{1,p}$-regularized maximum likelihood problem and discusses its properties. Section 2.4 describes our block coordinate descent method. Experimental results are in Section 2.5.

2.2 Background

In this chapter, we use the notation in Table 2.1.

A *Gaussian graphical model* can be represented by an undirected graph over N nodes, where the nodes correspond to continuous random variables which are jointly Gaussian. We will denote by $\mathbf{\Sigma} \in \mathbb{R}^{N \times N}$ the covariance matrix of the corresponding multivariate Gaussian distribution. Conditional independence in a Gaussian graphical model between two variables given the rest of the variables corresponds to zero entries in the *inverse covariance*, or *precision* matrix $\mathbf{\Omega} = \mathbf{\Sigma}^{-1}$ (15): namely, assuming $\mathbf{\Omega} = \{\omega_{n_1 n_2}\}$, two variables n_1 and n_2 are conditionally independent given the rest of the variables if and only if $\omega_{n_1 n_2} = 0$.

The estimation of sparse precision matrices was first introduced in (4). It is well known that finding the most sparse precision matrix which fits a dataset is a NP-hard problem (1). Therefore, several ℓ_1-regularization methods have been proposed.

Given a dense sample covariance matrix $\widehat{\mathbf{\Sigma}} \succeq \mathbf{0}$, the problem of finding a sparse precision matrix $\mathbf{\Omega}$ by regularized maximum likelihood estimation is given by:

$$\max_{\mathbf{\Omega} \succ \mathbf{0}} \left(\log \det \mathbf{\Omega} - \langle \widehat{\mathbf{\Sigma}}, \mathbf{\Omega} \rangle - \rho \|\mathbf{\Omega}\|_1 \right) \tag{2.1}$$

for $\rho > 0$. The term $\log \det \mathbf{\Omega} - \langle \widehat{\mathbf{\Sigma}}, \mathbf{\Omega} \rangle$ is the Gaussian log-likelihood. The term $\|\mathbf{\Omega}\|_1$ encourages sparseness of the precision matrix or conditional independence among variables.

Several algorithms have been proposed for solving eq.(2.1): *covariance selection* (1), *graphical lasso* (8) and the *Meinshausen-Bühlmann approximation* (25).

Besides sparseness, several regularizers have been proposed for Gaussian graphical models, for enforcing diagonal structure (19), spatial coherence (12), common structure among multiple tasks (13), or sparse changes in controlled experiments (39). In particular, different group sparse priors have been proposed for enforcing block structure for known block-variable assignments (5, 32) and unknown block-variable assignments (22, 23), or power law regularization in scale free networks (20).

Variable selection has been applied to very diverse problems, such as linear regression (33), classification (2, 18, 6) and reinforcement learning (28).

Structure learning through ℓ_1-regularization has been also proposed for different types of graphical models: Markov random fields (17); Bayesian networks on binary variables (31); Conditional random fields (30); and Ising models (36).

2.3 Preliminaries

In this section, we set up the problem and discuss some of its properties.

2.3.1 Problem Setup

We propose priors that are motivated from the variable selection literature from regression and classification, such as group lasso (37, 24, 27) which imposes an $\ell_{1,2}$-norm penalty, and simultaneous lasso (35, 34) which imposes an $\ell_{1,\infty}$-norm penalty.

Recall that an edge in a Gaussian graphical model corresponds to a non-zero entry in the precision matrix. We promote variable selection by learning a structure with a small number of nodes that interact with each other, or equivalently a large number of nodes that are disconnected from the rest of the graph. For each disconnected node, its corresponding row in the precision matrix (or column given that it is symmetric) contains only zeros (except for the diagonal). Therefore, the use of row-level regularizers such as the $\ell_{1,p}$-norm are natural in our context. Note that our goal differs from sparse Gaussian graphical models, in which sparseness is imposed at the edge level only. We additionally impose sparseness at the node level, which promotes conditional independence of variables with respect to all other variables.

Given a dense sample covariance matrix $\widehat{\boldsymbol{\Sigma}} \succeq \mathbf{0}$, we learn a precision matrix $\boldsymbol{\Omega} \in \mathbb{R}^{N \times N}$ for N variables. The *variable-selection structure learning problem* is defined as:

$$\max_{\boldsymbol{\Omega} \succ \mathbf{0}} \left(\log \det \boldsymbol{\Omega} - \langle \widehat{\boldsymbol{\Sigma}}, \boldsymbol{\Omega} \rangle - \rho \|\boldsymbol{\Omega}\|_1 - \tau \|\boldsymbol{\Omega}\|_{1,p} \right) \tag{2.2}$$

for $\rho > 0$, $\tau > 0$ and $p \in \{2, \infty\}$. The term $\log \det \boldsymbol{\Omega} - \langle \widehat{\boldsymbol{\Sigma}}, \boldsymbol{\Omega} \rangle$ is the Gaussian log-likelihood. $\|\boldsymbol{\Omega}\|_1$ encourages sparseness of the precision matrix or conditional independence among variables. The last term $\|\boldsymbol{\Omega}\|_{1,p}$ is our variable selection regularizer, and it is defined as:

$$\|\boldsymbol{\Omega}\|_{1,p} = \sum_n \|(\omega_{n,1}, \ldots, \omega_{n,n-1}, \omega_{n,n+1}, \ldots, \omega_{n,N})\|_p \tag{2.3}$$

In a technical report, (9) proposed an optimization problem that is similar to eq.(2.2). The main differences are that their model does not promote sparseness, and that they do not solve the original maximum likelihood problem, but instead build upon an approximation (pseudo-likelihood) approach of Meinshausen-Bühlmann (25) based on independent linear regression problems. Finally, note that regression based methods such as (25) have been already shown in (8) to have worse performance than solving the original maximum likelihood problem. In this work, we solve the original maximum likelihood problem.

2.3.2 Bounds

In what follows, we discuss uniqueness and boundedness of the optimal solution of our problem.

Lemma 2.1. *For $\rho > 0$, $\tau > 0$, the variable-selection structure learning problem in eq.(2.2) is a maximization problem with concave (but not strictly concave) objective function and convex constraints.*

Proof. The Gaussian log-likelihood is concave, since log det is concave on the space of symmetric positive definite matrices, and since the linear operator $\langle \cdot, \cdot \rangle$ is also concave. Both regularization terms, the negative ℓ_1-norm as well as the negative $\ell_{1,p}$-norm defined in eq.(2.3) are non-smooth concave functions. Finally, $\mathbf{\Omega} \succ \mathbf{0}$ is a convex constraint. $\square$

For clarity of exposition, we assume that the diagonals of $\mathbf{\Omega}$ are penalized by our variable selection regularizer defined in eq.(2.3).

Theorem 2.2. *For $\rho > 0$, $\tau > 0$, the optimal solution to the variable-selection structure learning problem in eq.(2.2) is unique and bounded as follows:*

$$\left(\frac{1}{\|\widehat{\mathbf{\Sigma}}\|_2 + N\rho + N^{1/p'}\tau} \right) \mathbf{I} \preceq \mathbf{\Omega}^* \preceq \left(\frac{N}{\max(\rho, \tau)} \right) \mathbf{I} \tag{2.4}$$

where $\ell_{p'}$-norm is the dual of the ℓ_p-norm, i.e. $(p = 2, p' = 2)$ or $(p = \infty, p' = 1)$.

Proof. By using the identity for dual norms $\kappa \|\mathbf{c}\|_p = \max_{\|\mathbf{d}\|_{p'} \leq \kappa} \mathbf{d}^{\mathrm{T}}\mathbf{c}$ in eq.(2.2), we get:

$$\max_{\mathbf{\Omega} \succ \mathbf{0}} \min_{\substack{\|\mathbf{A}\|_\infty \leq \rho \\ \|\mathcal{B}\|_{\infty,p'} \leq \tau}} \left(\log \det \mathbf{\Omega} - \langle \widehat{\mathbf{\Sigma}} + \mathbf{A} + \mathcal{B}, \mathbf{\Omega} \rangle \right) \tag{2.5}$$

where $\|B\|_{\infty,p'} = \max_n \|(b_{n,1}, \ldots, b_{n,N})\|_{p'}$. By virtue of Sion's minimax theorem, we can swap the order of max and min. Furthermore, note that the optimal solution of the inner equation is given by $\mathbf{\Omega} = (\widehat{\mathbf{\Sigma}} + \mathbf{A} + \mathcal{B})^{-1}$. By replacing this solution in eq.(2.5), we get the dual problem of eq.(2.2):

$$\min_{\substack{\|\mathbf{A}\|_\infty \leq \rho \\ \|\mathcal{B}\|_{\infty,p'} \leq \tau}} \left(-\log \det(\widehat{\mathbf{\Sigma}} + \mathbf{A} + \mathcal{B}) - N \right) \tag{2.6}$$

In order to find a lower bound for the minimum eigenvalue of $\mathbf{\Omega}^*$, note that $\|\mathbf{\Omega}^{*-1}\|_2 = \|\widehat{\mathbf{\Sigma}} + \mathbf{A} + \mathcal{B}\|_2 \leq \|\widehat{\mathbf{\Sigma}}\|_2 + \|\mathbf{A}\|_2 + \|\mathcal{B}\|_2 \leq \|\widehat{\mathbf{\Sigma}}\|_2 + N\|\mathbf{A}\|_\infty + N^{1/p'}\|\mathcal{B}\|_{\infty,p'} \leq \|\widehat{\mathbf{\Sigma}}\|_2 + N\rho + N^{1/p'}\tau$. (Here we used $\|\mathcal{B}\|_2 \leq N^{1/p'}\|\mathcal{B}\|_{\infty,p'}$

as shown in Appendix 2.7)

In order to find an upper bound for the maximum eigenvalue of $\boldsymbol{\Omega}^*$, note that, at optimum, the primal-dual gap is zero:

$$-N + \langle \widehat{\boldsymbol{\Sigma}}, \boldsymbol{\Omega}^* \rangle + \rho \|\boldsymbol{\Omega}^*\|_1 + \tau \|\boldsymbol{\Omega}^*\|_{1,p} = 0 \qquad (2.7)$$

The upper bound is found as follows: $\|\boldsymbol{\Omega}^*\|_2 \leq \|\boldsymbol{\Omega}^*\|_{\mathfrak{F}} \leq \|\boldsymbol{\Omega}^*\|_1 = (N - \langle \widehat{\boldsymbol{\Sigma}}, \boldsymbol{\Omega}^* \rangle - \tau \|\boldsymbol{\Omega}^*\|_{1,p})/\rho$. Note that $\tau \|\boldsymbol{\Omega}^*\|_{1,p} \geq 0$, and since $\widehat{\boldsymbol{\Sigma}} \succeq \mathbf{0}$ and $\boldsymbol{\Omega}^* \succ \mathbf{0}$, it follows that $\langle \widehat{\boldsymbol{\Sigma}}, \boldsymbol{\Omega}^* \rangle \geq 0$. Therefore, $\|\boldsymbol{\Omega}^*\|_2 \leq \frac{N}{\rho}$. In a similar fashion, $\|\boldsymbol{\Omega}^*\|_2 \leq \|\boldsymbol{\Omega}^*\|_{1,p} = (N - \langle \widehat{\boldsymbol{\Sigma}}, \boldsymbol{\Omega}^* \rangle - \rho \|\boldsymbol{\Omega}^*\|_1)/\tau$. (Here we used $\|\boldsymbol{\Omega}^*\|_2 \leq \|\boldsymbol{\Omega}^*\|_{1,p}$ as shown in Appendix 2.7). Note that $\rho \|\boldsymbol{\Omega}^*\|_1 \geq 0$ and $\langle \widehat{\boldsymbol{\Sigma}}, \boldsymbol{\Omega}^* \rangle \geq 0$. Therefore, $\|\boldsymbol{\Omega}^*\|_2 \leq \frac{N}{\tau}$.   □

2.4   Block Coordinate Descent Method

Since the objective function in eq.(2.2) contains a non-smooth regularizer, methods such as gradient descent cannot be applied. On the other hand, subgradient descent methods very rarely converge to non-smooth points (7). In our problem, these non-smooth points correspond to zeros in the precision matrix, are often the true minima of the objective function, and are very desirable in the solution because they convey information of conditional independence among variables.

We apply block coordinate descent method on the primal problem (12, 13), unlike covariance selection (1) and graphical lasso (8) which optimize the dual. Optimization of the dual problem in eq.(2.6) by a block coordinate descent method can be done with quadratic programming for $p = \infty$ but not for $p = 2$ (i.e. the objective function is quadratic for $p \in \{2, \infty\}$, the constraints are linear for $p = \infty$ and quadratic for $p = 2$). Optimization of the primal problem provides the same efficient framework for $p \in \{2, \infty\}$. We point out that a projected subgradient method as in (5) cannot be applied since our regularizer does not decompose into disjoint subsets. Our problem contains a positive definiteness constraint and therefore it does not fall in the general framework of (37, 24, 27, 29, 35, 34) which consider unconstrained problems only. Finally, more recent work of (3, 21) consider subsets with overlap, but it does still consider unconstrained problems only.

Theorem 2.3. *The block coordinate descent method for the variable-selection structure learning problem in eq.(2.2) generates a sequence of positive definite solutions.*

Proof. Maximization can be performed with respect to one row and column of all precision matrices $\mathbf{\Omega}$ at a time. Without loss of generality, we use the last row and column in our derivation. Let:

$$\mathbf{\Omega} = \begin{bmatrix} \mathbf{W} & \mathbf{y} \\ \mathbf{y}^{\mathrm{T}} & z \end{bmatrix} \quad , \quad \widehat{\mathbf{\Sigma}} = \begin{bmatrix} \mathbf{S} & \mathbf{u} \\ \mathbf{u}^{\mathrm{T}} & v \end{bmatrix} \tag{2.8}$$

where $\mathbf{W}, \mathbf{S} \in \mathbb{R}^{N-1 \times N-1}$, $\mathbf{y}, \mathbf{u} \in \mathbb{R}^{N-1}$.

In terms of the variables $\mathbf{y}, z$ and the constant matrix $\mathbf{W}$, the variable-selection structure learning problem in eq.(2.2) can be reformulated as:

$$\max_{\mathbf{\Omega} \succ \mathbf{0}} \begin{pmatrix} \log(z - \mathbf{y}^{\mathrm{T}} \mathbf{W}^{-1} \mathbf{y}) - 2\mathbf{u}^{\mathrm{T}} \mathbf{y} - (v + \rho)z \\ -2\rho \|\mathbf{y}\|_1 - \tau \|\mathbf{y}\|_p - \tau \sum_n \|(y_n, t_n)\|_p \end{pmatrix} \tag{2.9}$$

where $t_n = \|(w_{n,1}, \ldots, w_{n,n-1}, w_{n,n+1}, \ldots, w_{n,N})\|_p$.

If $\mathbf{\Omega}$ is a symmetric matrix, according to the Haynsworth inertia formula, $\mathbf{\Omega} \succ \mathbf{0}$ if and only if its Schur complement $z - \mathbf{y}^{\mathrm{T}} \mathbf{W}^{-1} \mathbf{y} > 0$ and $\mathbf{W} \succ \mathbf{0}$. By maximizing eq.(2.9) with respect to z, we get:

$$z - \mathbf{y}^{\mathrm{T}} \mathbf{W}^{-1} \mathbf{y} = \frac{1}{v + \rho} \tag{2.10}$$

and since $v > 0$ and $\rho > 0$, this implies that the Schur complement in eq.(2.10) is positive. Finally, in our iterative optimization, it suffices to initialize $\mathbf{\Omega}$ to a matrix known to be positive definite, e.g. a diagonal matrix with positive elements. $\qquad \square$

Theorem 2.4. *The block coordinate descent method for the variable-selection structure learning problem in eq.(2.2) is equivalent to solving a sequence of strictly convex $(\ell_1, \ell_{1,p})$ regularized quadratic subproblems for $p \in \{2, \infty\}$:*

$$\min_{\mathbf{y} \in \mathbb{R}^{N-1}} \begin{pmatrix} \frac{1}{2} \mathbf{y}^{\mathrm{T}} (v + \rho) \mathbf{W}^{-1} \mathbf{y} + \mathbf{u}^{\mathrm{T}} \mathbf{y} \\ +\rho \|\mathbf{y}\|_1 + \frac{\tau}{2} \|\mathbf{y}\|_p + \frac{\tau}{2} \sum_n \|(y_n, t_n)\|_p \end{pmatrix} \tag{2.11}$$

Proof. By replacing the optimal z given by eq.(2.10) into the objective function in eq.(2.9), we get eq.(2.11). Since $\mathbf{W} \succ \mathbf{0} \Rightarrow \mathbf{W}^{-1} \succ \mathbf{0}$, hence eq.(2.11) is strictly convex. $\qquad \square$

Lemma 2.5. *If $\|\mathbf{u}\|_\infty \leq \rho + \tau/(2(N-1)^{1/p'})$ or $\|\mathbf{u}\|_{p'} \leq \rho + \tau/2$, the $(\ell_1, \ell_{1,p})$ regularized quadratic problem in eq.(2.11) has the minimizer $\mathbf{y}^* = \mathbf{0}$.*

Proof. Note that since $\mathbf{W} \succ \mathbf{0} \Rightarrow \mathbf{W}^{-1} \succ \mathbf{0}$, $\mathbf{y}^* = \mathbf{0}$ is the minimizer of the quadratic part of eq.(2.11). It suffices to prove that the remaining part is also minimized for $\mathbf{y}^* = \mathbf{0}$, i.e. $\mathbf{u}^{\mathrm{T}} \mathbf{y} + \rho \|\mathbf{y}\|_1 + \frac{\tau}{2} \|\mathbf{y}\|_p + \frac{\tau}{2} \sum_n \|(y_n, t_n)\|_p \geq \frac{\tau}{2} \sum_n t_n$ for an arbitrary $\mathbf{y}$. The lower bound comes from setting $\mathbf{y}^* = \mathbf{0}$ in eq.(2.11) and by noting that $(\forall n)\ t_n > 0$.

By using lower bounds $\sum_n \|(y_n, t_n)\|_p \geq \sum_n t_n$ and either $\|\mathbf{y}\|_p \geq \|\mathbf{y}\|_1/(N-1)^{1/p'}$ or $\|\mathbf{y}\|_1 \geq \|\mathbf{y}\|_p$, we modify the original claim into a stronger one, i.e. $\mathbf{u}^T\mathbf{y} + (\rho + \tau/(2(N-1)^{1/p'}))\|\mathbf{y}\|_1 \geq 0$ or $\mathbf{u}^T\mathbf{y} + (\rho + \tau/2)\|\mathbf{y}\|_p \geq 0$. Finally, by using the identity for dual norms $\kappa\|\mathbf{y}\|_p = \max_{\|\mathbf{d}\|_{p'} \leq \kappa} \mathbf{d}^T\mathbf{y}$, we have $\max_{\|\mathbf{d}\|_\infty \leq \rho + \tau/(2(N-1)^{1/p'})} (\mathbf{u} + \mathbf{d})^T\mathbf{y} \geq 0$ or $\max_{\|\mathbf{d}\|_{p'} \leq \rho + \tau/2} (\mathbf{u} + \mathbf{d})^T\mathbf{y} \geq 0$, which proves our claim. $\qquad\Box$

Remark 2.6. *By using Lemma 2.5, we can reduce the size of the original problem by removing variables in which this condition holds, since it only depends on the dense sample covariance matrix.*

Theorem 2.7. *The coordinate descent method for the $(\ell_1, \ell_{1,p})$ regularized quadratic problem in eq.(2.11) is equivalent to solving a sequence of strictly convex (ℓ_1, ℓ_p) regularized quadratic subproblems:*

$$\min_x \left(\frac{1}{2}qx^2 - cx + \rho|x| + \frac{\tau}{2}\|(x, a)\|_p + \frac{\tau}{2}\|(x, b)\|_p \right) \qquad (2.12)$$

Proof. Without loss of generality, we use the last row and column in our derivation, since permutation of rows and columns is always possible. Let:

$$\mathbf{W}^{-1} = \begin{bmatrix} \mathcal{H}_{11} & \mathbf{h}_{12} \\ \mathbf{h}_{12}{}^T & h_{22} \end{bmatrix} \quad, \quad \mathbf{y} = \begin{bmatrix} \mathbf{y}_1 \\ x \end{bmatrix} \quad, \quad \mathbf{u} = \begin{bmatrix} \mathbf{u}_1 \\ u_2 \end{bmatrix} \qquad (2.13)$$

where $\mathcal{H}_{11} \in \mathbb{R}^{N-2 \times N-2}$, $\mathbf{h}_{12}, \mathbf{y}_1, \mathbf{u}_1 \in \mathbb{R}^{N-2}$.

In terms of the variable x and the constants $q = (v + \rho)h_{22}$, $c = -((v+\rho)\mathbf{h}_{12}{}^T\mathbf{y}_1 + u_2)$, $a = \|\mathbf{y}_1\|_p$, $b = t_n$, the $(\ell_1, \ell_{1,p})$ regularized quadratic problem in eq.(2.11) can be reformulated as in eq.(2.12). Moreover, since $v > 0 \wedge \rho > 0 \wedge h_{22} > 0 \Rightarrow q > 0$, and therefore eq.(2.12) is strictly convex. $\qquad\Box$

For $p = \infty$, eq.(2.12) has five points in which the objective function is non-smooth, i.e. $x \in \{-\max(a, b), -\min(a, b), 0, \min(a, b), \max(a, b)\}$. Furthermore, since the objective function is quadratic on each interval, it admits a closed form solution.

For $p = 2$, eq.(2.12) has only one non-smooth point, i.e. $x = 0$. Given the objective function $f(x)$, we first compute the left derivative $\partial_- f(0) = -c - \rho$ and the right derivative $\partial_+ f(0) = -c + \rho$. If $\partial_- f(0) \leq 0 \wedge \partial_+ f(0) \geq 0 \Rightarrow x^* = 0$. If $\partial_- f(0) > 0 \Rightarrow x^* < 0$ and we use the one-dimensional *Newton-Raphson method* for finding x^*. If $\partial_+ f(0) < 0 \Rightarrow x^* > 0$. For numerical stability, we add a small $\varepsilon > 0$ to the ℓ_2-norms by using $\sqrt{x^2 + a^2 + \varepsilon}$ instead of $\|(x, a)\|_2$.

Algorithm 2.1 shows the block coordinate descent method in detail. A careful implementation leads to a time complexity of $\mathcal{O}(KN^3)$ for K iterations and N variables. In our experiments, the algorithm converges quickly

---

Algorithm 2.1 Block Coordinate Descent

Input: $\widehat{\boldsymbol{\Sigma}} \succeq \mathbf{0}$, $\rho > 0$, $\tau > 0$, $p \in \{2, \infty\}$

Initialize $\boldsymbol{\Omega} = \text{diag}(\widehat{\boldsymbol{\Sigma}})^{-1}$

for each iteration $1, \dots, K$ and each variable $1, \dots, N$ **do**

    Split $\boldsymbol{\Omega}$ into $\mathbf{W}, \mathbf{y}, z$ and $\widehat{\boldsymbol{\Sigma}}$ into $\mathbf{S}, \mathbf{u}, v$ as described in eq.(2.8)

    Update $\mathbf{W}^{-1}$ by using the Sherman-Woodbury-Morrison formula (Note that when iterating from one variable to the next one, only one row and column change on matrix $\mathbf{W}$)

    for each variable $1, \dots, N-1$ **do**

        Split $\mathbf{W}^{-1}, \mathbf{y}, \mathbf{u}$ as in eq.(2.13)

        Solve the (ℓ_1, ℓ_p) regularized quadratic problem in closed form ($p = \infty$) or by using the Newton-Raphson method ($p = 2$)

    end for

    Update $z \leftarrow \frac{1}{v+\rho} + \mathbf{y}^{\mathsf{T}} \mathbf{W}^{-1} \mathbf{y}$

end for

Output: $\boldsymbol{\Omega} \succ \mathbf{0}$

---

in usually $K = 10$ iterations. Polynomial dependence $\mathcal{O}(N^3)$ on the number of variables is expected since no algorithm can be faster than computing the inverse of the sample covariance in the case of an infinite sample.

2.5   Experimental Results

We test with a synthetic example the ability of the method to recover ground truth structure from data. The model contains $N \in \{50, 100, 200\}$ variables. For each of 50 repetitions, we first select a proportion of "connected" nodes (either 0.2,0.5,0.8) from the N variables. The unselected (i.e. "disconnected") nodes do not participate in any edge of the ground truth model. We then generate edges among the connected nodes with a required density (either 0.2,0.5,0.8), where each edge weight is generated uniformly at random from $\{-1, +1\}$. We ensure positive definiteness of $\boldsymbol{\Omega}_g$ by verifying that its minimum eigenvalue is at least 0.1. We then generate a dataset of 50 samples. We model the ratio $\bar{\sigma}_c/\bar{\sigma}_d$ between the standard deviation of connected versus disconnected nodes. In the "high variance confounders" regime, $\bar{\sigma}_c/\bar{\sigma}_d = 1$ which means that on average connected and disconnected variables have the same standard deviation. In the "low variance confounders" regime, $\bar{\sigma}_c/\bar{\sigma}_d = 10$ which means that on average the standard deviation of a connected variable is 10 times the one of a disconnected variable. Variables with low variance produce higher values in the precision matrix than variables with high variance. We analyze both regimes in order to evaluate the impact of this effect in structure recovery.

In order to measure the closeness of the recovered models to the ground truth, we measured the Kullback-Leibler (KL) divergence, sensitivity (one minus the fraction of falsely excluded edges) and specificity (one minus the fraction of falsely included edges). We compare to the following methods: covariance selection (1), graphical lasso (8), Meinshausen-Bühlmann approximation (25) and Tikhonov regularization. For our method, we found that the variable selection parameter $\tau = 50\rho$ provides reasonable results, in both synthetic and real-world experiments. Therefore, we report results only with respect to the sparseness parameter ρ.

First, we test the performance of our methods for increasing number of variables, moderate edge density (0.5) and high proportion of connected nodes (0.8). Figure 2.1 and Figure 2.2 show the ROC curves and KL divergence between the recovered models and the ground truth. In both "low" and "high variance confounders" regimes, our $\ell_{1,2}$ and $\ell_{1,\infty}$ methods recover ground truth edges better than competing methods (higher ROC) and produce better probability distributions (lower KL divergence) than the other methods. Our methods degrade less than competing methods in recovering the ground truth edges when the number of variables grows, while the KL divergence behavior remains similar.

Second, we test the performance of our methods with respect to edge density and the proportion of connected nodes. Figure 2.3 shows the KL divergence between the recovered models and the ground truth for the "low variance confounders" regime. Our $\ell_{1,2}$ and $\ell_{1,\infty}$ methods produce better probability distributions (lower KL divergence) than the remaining techniques. (Please, see Appendix 2.8 for results on ROC and the "high variance confounders" regime.)

Our $\ell_{1,2}$ method takes 0.07s for $N = 100$, 0.12s for $N = 200$ variables. Our $\ell_{1,\infty}$ method takes 0.13s for $N = 100$, 0.63s for $N = 200$. Graphical lasso (8), the fastest and most accurate competing method in our evaluation, takes 0.11s for $N = 100$, 0.49s for $N = 200$. Our $\ell_{1,\infty}$ method is slightly slower than graphical lasso, while our $\ell_{1,2}$ method is the fastest. One reason for this is that Lemma 2.5 eliminates more variables in the $\ell_{1,2}$ setting.

For experimental validation on real-world datasets, we use datasets with a diverse nature of probabilistic relationships: brain fMRI, gene expression, NASDAQ stock prices and world weather. The *brain fMRI* dataset collected by (11) captures brain function of 15 cocaine addicted and 11 control subjects under conditions of monetary reward. Each subject contains 87 scans of $53 \times 63 \times 46$ voxels each, taken every 3.5 seconds. Registration to a common spatial template and spatial smoothing was done in SPM2 (`http://www.fil.ion.ucl.ac.uk/spm/`). After sampling each $4 \times 4 \times 4$ voxels, we obtained 869 variables. The *gene expression* dataset contains 8,565 variables and 587 samples. The dataset was collected by (26) from drug treated rat

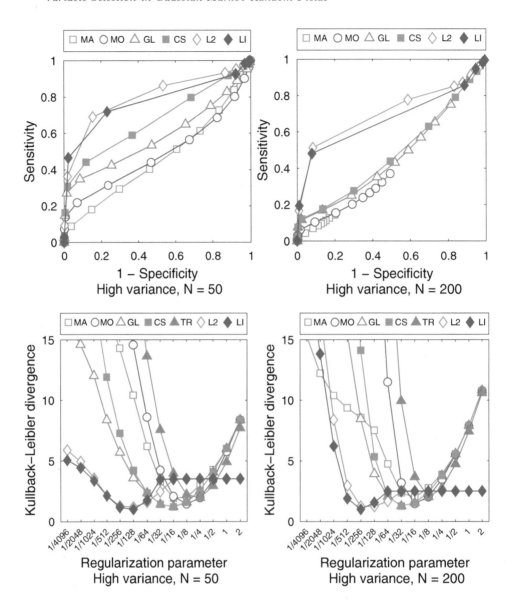

Figure 2.1: ROC curves (first row) and KL divergence (second row) for the "high variance confounders" regime. Left: $N = 50$ variables, right: $N = 200$ variables (connectedness 0.8, edge density 0.5). Our proposed methods $\ell_{1,2}$ (L2) and $\ell_{1,\infty}$ (LI) recover edges better and produce better probability distributions than Meinshausen-Bühlmann with AND-rule (MA), OR-rule (MO), graphical lasso (GL), covariance selection (CS) and Tikhonov regularization (TR). Our methods degrade less in recovering the ground truth edges when the number of variables grows.

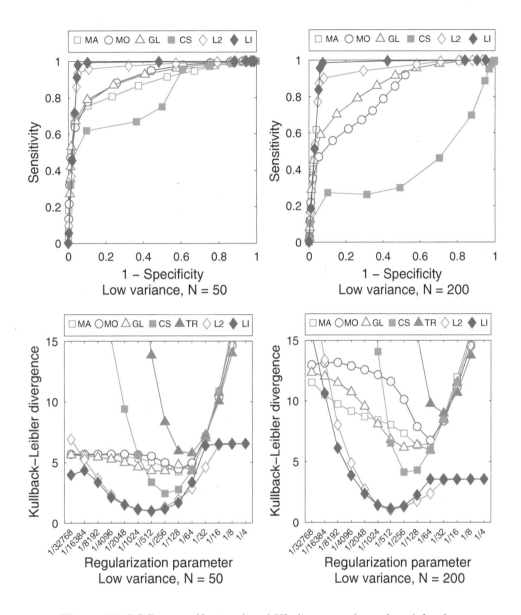

Figure 2.2: ROC curves (first row) and KL divergence (second row) for the "low variance confounders" regime. Left: $N = 50$ variables, right: $N = 200$ variables (connectedness 0.8, edge density 0.5). Our proposed methods $\ell_{1,2}$ (L2) and $\ell_{1,\infty}$ (LI) recover edges better and produce better probability distributions than Meinshausen-Bühlmann with AND-rule (MA), OR-rule (MO), graphical lasso (GL), covariance selection (CS) and Tikhonov regularization (TR). Our methods degrade less in recovering the ground truth edges when the number of variables grows.

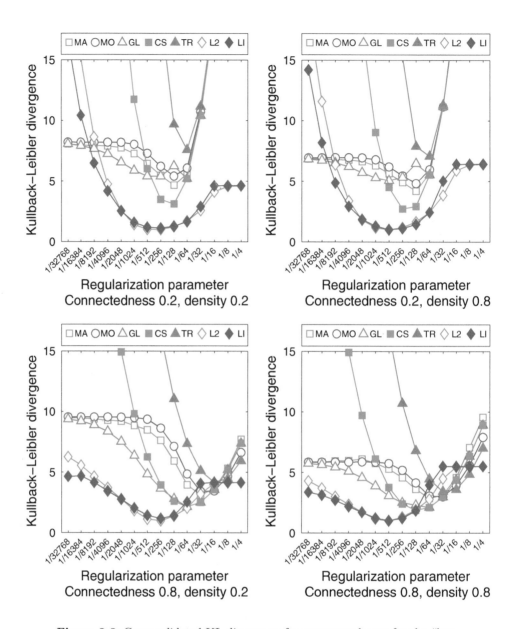

Figure 2.3: Cross-validated KL divergence for structures learnt for the "low variance confounders" regime ($N = 50$ variables, different connectedness and density levels). Our proposed methods $\ell_{1,2}$ (L2) and $\ell_{1,\infty}$ (LI) produce better probability distributions than Meinshausen-Bühlmann with AND-rule (MA), OR-rule (MO), graphical lasso (GL), covariance selection (CS) and Tikhonov regularization (TR).

livers, by treating rats with a variety of fibrate, statin, or estrogen receptor agonist compounds. The dataset is publicly available at `http://www.ebi.ac.uk/`. In order to consider the full set of genes, we had to impute a very small percentage (0.90%) of missing values by randomly generating values with the same mean and standard deviation. The *NASDAQ stocks* dataset contains daily opening and closing prices for 2,749 stocks from Apr 19, 2010 to Apr 18, 2011 (257 days). The dataset was downloaded from `http://www.google.com/finance`. For our experiments, we computed the percentage of change between the closing and opening prices. The *world weather* dataset contains monthly measurements of temperature, precipitation, vapor, cloud cover, wet days and frost days from Jan 1990 to Dec 2002 (156 months) on a 2.5×2.5 degree grid that covers the entire world. The dataset is publicly available at `http://www.cru.uea.ac.uk/`. After sampling each 5×5 degrees, we obtained 4,146 variables. For our experiments, we computed the change between each month and the month in the previous year.

For all the datasets, we used one third of the data for training, one third for validation and the remaining third for testing. Since the brain fMRI dataset has a very small number of subjects, we performed six repetitions by making each third of the data take turns as training, validation and testing sets. In our evaluation, we included scale free networks (20). We did not include the covariance selection method (1) since we found it is extremely slow for these high-dimensional datasets. We report the negative log-likelihood on the testing set in Figure 2.4 (we subtracted the entropy measured on the testing set and then scaled the results for visualization purposes). We can observe that the log-likelihood of our method is remarkably better than the other techniques for all the datasets.

Regarding comparison to group sparse methods, in our previous experiments we did not include block structure for known block-variable assignments (5, 32) since our synthetic and real-world datasets lack such assignments. We did not include block structure for unknown assignments (22, 23) given their time complexity ((23) has a $\mathcal{O}(N^5)$-time Gibbs sampler step for N variables and it is applied for $N = 60$ only, while (22) has a $\mathcal{O}(N^4)$-time ridge regression step). Instead, we evaluated our method in the *baker's yeast* gene expression dataset in (5) which contains 677 variables and 173 samples. We used the experimental settings of Fig.3 in (22). For learning one structure, (22) took 5 hours while our $\ell_{1,2}$ method took only 50 seconds. Our method outperforms block structures for known and unknown assignments. The log-likelihood is 0 for Tikhonov regularization, 6 for (5, 22), 8 for (32), and 22 for our $\ell_{1,2}$ method.

We show the structures learnt for cocaine addicted and control subjects in Figure 2.5, for our $\ell_{1,2}$ method and graphical lasso (8). The disconnected

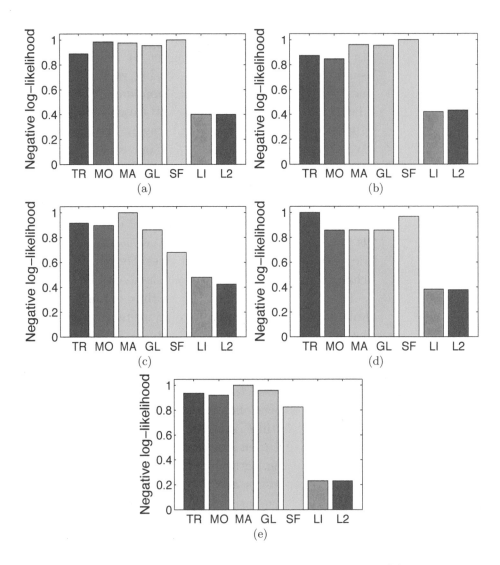

Figure 2.4: Test negative log-likelihood of structures learnt for (a) addicted subjects and (b) control subjects in the brain fMRI dataset, (c) gene expression, (d) NASDAQ stocks and (e) world weather. Our proposed methods $\ell_{1,2}$ (L2) and $\ell_{1,\infty}$ (LI) outperforms the Meinshausen-Bühlmann with AND-rule (MA), OR-rule (MO), graphical lasso (GL), Tikhonov regularization (TR) and scale free networks (SF).

variables are not shown. Note that our structures involve remarkably fewer connected variables but yield a higher log-likelihood than graphical lasso (Figure 2.4), which suggests that the discarded edges from the disconnected nodes are not important for accurate modeling of this dataset. Moreover, removal of a large number of nuisance variables (voxels) results into a more interpretable model, clearly demonstrating brain areas involved in structural model differences that discriminate cocaine addicted from control subjects. Note that graphical lasso (bottom of Figure 2.5) connects most of the brain voxels in both populations, making them impossible to compare. Our approach produces more "localized" networks (top of the Figure 2.5) involving a relatively small number of brain areas: cocaine addicted subjects show increased interactions between the visual cortex (back of the brain, on the left in the image) and the prefrontal cortex (front of the brain, on the right in the image), while at the same time decreased density of interactions between the visual cortex with other brain areas (more clearly present in control subjects). The alteration in this pathway in the addict group is highly significant from a neuroscientific perspective. First, the trigger for reward was a visual stimulus. Abnormalities in the visual cortex was reported in (16) when comparing cocaine abusers to control subjects. Second, the prefrontal cortex is involved in higher-order cognitive functions such as decision making and reward processing. Abnormal monetary processing in the prefrontal cortex was reported in (10) when comparing cocaine addicted individuals to controls. Although a more careful interpretation of the observed results remains to be done in the near future, these results are encouraging and lend themselves to specific neuroscientific hypothesis testing.

In a different evaluation, we used generatively learnt structures for a classification task. We performed a five-fold cross-validation on the subjects. From the subjects in the training set, we learned one structure for cocaine addicted and one structure for control subjects. Then, we assigned a test subject to the structure that gave highest probability for his data. All methods in our evaluation except Tikhonov regularization obtained 84.6% accuracy. Tikhonov regularization obtained 65.4% accuracy. Therefore, our method produces structures that retain discriminability with respect to standard sparseness promoting techniques.

2.6   Conclusions and Future Work

In this chapter, we presented variable selection in the context of learning sparse Gaussian graphical models by adding an $\ell_{1,p}$-norm regularization term, for $p \in \{2, \infty\}$. We presented a block coordinate descent method which yields sparse and positive definite estimates. We solved the original

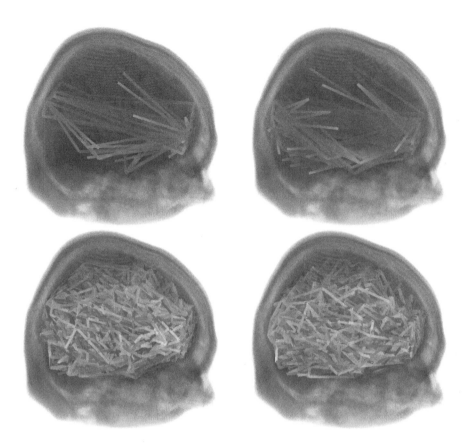

Figure 2.5: Structures learnt for cocaine addicted (left) and control subjects (right), for our $\ell_{1,2}$ method (top) and graphical lasso (bottom). Regularization parameter $\rho = 1/16$. Positive interactions in blue, negative interactions in red. Our structures are sparser (density 0.0016) than graphical lasso (density 0.023) where the number of edges in a complete graph is approximately 378000.

problem by efficiently solving a sequence of strictly convex (ℓ_1, ℓ_p) regularized quadratic minimization subproblems.

The motivation behind this work was to incorporate variable selection into structure learning of sparse Markov networks, and specifically Gaussian graphical models. Besides providing a better regularizer (as observed on several real-world datasets: brain fMRI, gene expression, NASDAQ stock prices and world weather), key advantages of our approach include a more accurate structure recovery in the presence of multiple noisy variables (as demonstrated by simulations), significantly better interpretability and same discriminability of the resulting network in practical applications (as shown for brain fMRI analysis).

There are several ways to extend this research. In practice, our technique converges in a small number of iterations, but an analysis of convergence rate needs to be performed. Consistency when the number of samples grows to infinity needs to be proved.

Acknowledgments

We thank Rita Goldstein for providing us the fMRI dataset. This work was supported in part by NIH Grants 1 R01 DA020949 and 1 R01 EB007530.

2.7   Appendix: Technical Lemma

In Theorem 2.2, we use four matrix norm inequalities that are less common in the literature. In this section, we prove them in detail.

Lemma 2.8. *For* $\mathbf{A} \in \mathbb{R}^{N \times N}$, *the following conditions hold:*

i. $\|\mathbf{A}\|_2 \leq \sqrt{N}\|\mathbf{A}\|_{\infty,2}$

ii. $\|\mathbf{A}\|_2 \leq N\|\mathbf{A}\|_{\infty,1}$

iii. $\|\mathbf{A}\|_2 \leq \|\mathbf{A}\|_{1,2}$

iv. $\mathbf{A} \succ \mathbf{0} \Rightarrow \|\mathbf{A}\|_2 \leq \|\mathbf{A}\|_{1,\infty}$

(2.14)

Proof. Claim i follows from $\|\mathbf{A}\|_2 \leq \|\mathbf{A}\|_{\mathfrak{F}} \leq \sqrt{N}\|\mathbf{A}\|_{\infty,2}$. The last inequality is equivalent to $\|\mathbf{A}\|_{\mathfrak{F}}^2 \leq N\|\mathbf{A}\|_{\infty,2}^2 \Rightarrow \sum_{n_1 n_2} a_{n_1 n_2}^2 \leq N \max_{n_1} (\sum_{n_2} a_{n_1 n_2}^2)$. Let $|c_{n_1}| = \sum_{n_2} a_{n_1 n_2}^2$, we get $\sum_{n_1} |c_{n_1}| \leq N \max_{n_1} |c_{n_1}|$. This is equivalent to $\|\mathbf{c}\|_1 \leq N\|\mathbf{c}\|_\infty$, and we prove our claim.

Claim ii follows from $\|\mathbf{A}\|_2 \leq \|\mathbf{A}\|_{\mathfrak{F}} \leq \|\mathbf{A}\|_1 \leq N\|\mathbf{A}\|_{\infty,1}$. The last inequality is equivalent to $\sum_{n_1 n_2} |a_{n_1 n_2}| \leq N \max_{n_1} (\sum_{n_2} |a_{n_1 n_2}|)$. Let $|c_{n_1}| =$

$\sum_{n_2} |a_{n_1 n_2}|$, we get $\sum_{n_1} |c_{n_1}| \leq N \max_{n_1} |c_{n_1}|$. This is equivalent to $\|\mathbf{c}\|_1 \leq N \|\mathbf{c}\|_\infty$, and we prove our claim.

Claim iii follows from $\|\mathbf{A}\|_2 \leq \|\mathbf{A}\|_{\mathfrak{F}} \leq \|\mathbf{A}\|_{1,2}$. The last inequality is equivalent to $\sqrt{\sum_{n_1 n_2} a_{n_1 n_2}^2} \leq \sum_{n_1} \sqrt{\sum_{n_2} a_{n_1 n_2}^2}$. Let $c_{n_1}^2 = \sum_{n_2} a_{n_1 n_2}^2$, we get $\sqrt{\sum_{n_1} c_{n_1}^2} \leq \sum_{n_1} \sqrt{c_{n_1}^2} = \sum_{n_1} |c_{n_1}|$. This is equivalent to $\|\mathbf{c}\|_2 \leq \|\mathbf{c}\|_1$, and we prove our claim.

Claim iv further assumes that $\mathbf{A}$ is symmetric and positive definite. In this case the spectral radius is less than or equal to any induced norm, specifically the $\ell_{\infty,1}$-norm also called the *max absolute row sum* norm. The inequality we want to prove is $\|\mathbf{A}\|_2 \leq \|\mathbf{A}\|_{\infty,1} \leq \|\mathbf{A}\|_{1,\infty}$. The last inequality is equivalent to $\max_{n_1} \left(\sum_{n_2} |a_{n_1 n_2}| \right) \leq \sum_{n_1} \left(\max_{n_2} |a_{n_1 n_2}| \right)$, which follows from the Jensen's inequality. □

2.8   Appendix: Additional Experimental Results

In what follows, we test the performance of our methods with respect to edge density and the proportion of connected nodes. The following results complement Figure 2.3 which reported KL divergence between the recovered models and the ground truth for the "low variance confounders" regime. Figure 2.6 and Figure 2.7 show the ROC curves and KL divergence between the recovered models and the ground truth for the "high variance confounders" regime. Our $\ell_{1,2}$ and $\ell_{1,\infty}$ methods recover ground truth edges better than competing methods (higher ROC) when edge density among connected nodes is high (0.8), regardless of the proportion of connected nodes. Our proposed methods get similarly good probability distributions (comparable KL divergence) than the other techniques. In the "low variance confounders" regime reported in Figure 2.8 and Figure 2.3, our proposed methods produce better probability distributions (lower KL divergence) than the remaining techniques. The behavior of the ROC curves is similar to the "high variance confounders" regime.

References

1. O. Banerjee, L. El Ghaoui, A. d'Aspremont, and G. Natsoulis. Convex optimization techniques for fitting sparse Gaussian graphical models. *ICML*, 2006.

2. A. Chan, N. Vasconcelos, and G. Lanckriet. Direct convex relaxations of sparse SVM. *ICML*, 2007.

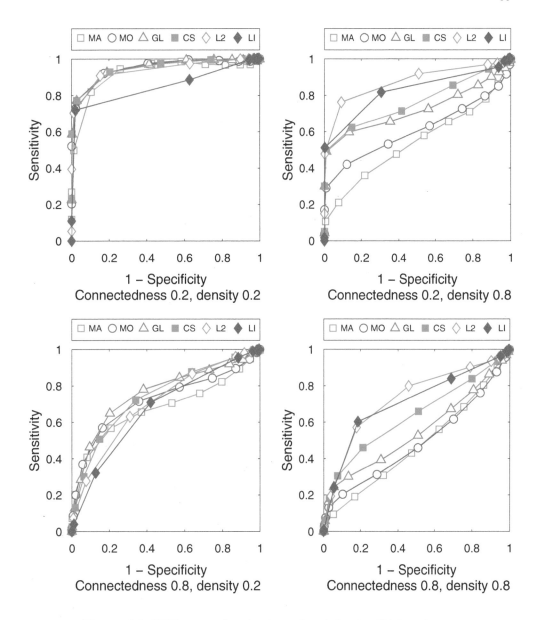

Figure 2.6: ROC curves for structures learnt for the "high variance confounders" regime ($N = 50$ variables, different connectedness and density levels). Our proposed methods $\ell_{1,2}$ (L2) and $\ell_{1,\infty}$ (LI) recover the ground truth edges better than Meinshausen-Bühlmann with AND-rule (MA), OR-rule (MO), graphical lasso (GL) and covariance selection (CS), when the edge density among the connected nodes is high. (See charts on the right.)

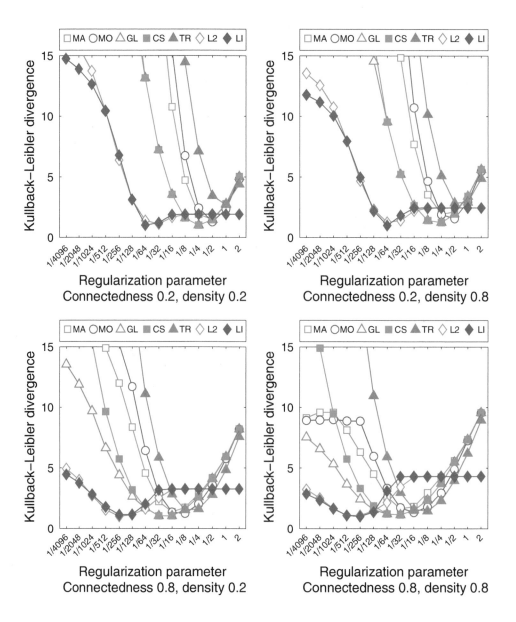

Figure 2.7: Cross-validated KL divergence for structures learnt for the "high variance confounders" regime ($N = 50$ variables, different connectedness and density levels). Our proposed methods $\ell_{1,2}$ (L2) and $\ell_{1,\infty}$ (LI) produce similarly good probability distributions than Meinshausen-Bühlmann with AND-rule (MA), OR-rule (MO), graphical lasso (GL), covariance selection (CS) and Tikhonov regularization (TR).

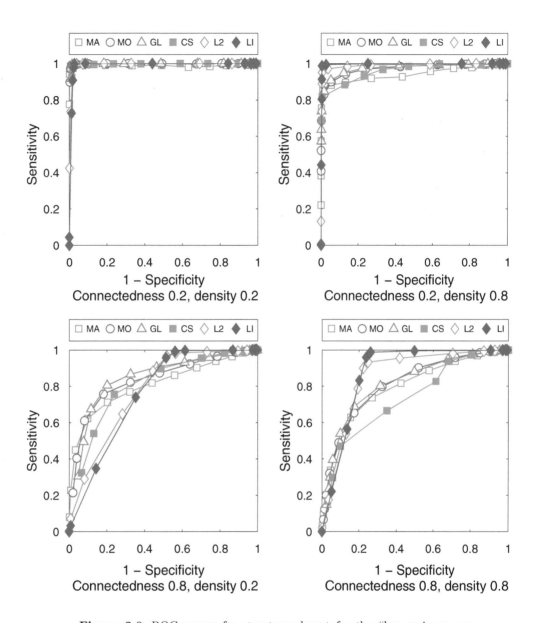

Figure 2.8: ROC curves for structures learnt for the "low variance confounders" regime ($N = 50$ variables, different connectedness and density levels). Our proposed methods $\ell_{1,2}$ (L2) and $\ell_{1,\infty}$ (LI) recover the ground truth edges better than Meinshausen-Bühlmann with AND-rule (MA), OR-rule (MO), graphical lasso (GL) and covariance selection (CS), when the edge density among the connected nodes is high. (See charts on the right.)

3. X. Chen, Q. Lin, S. Kim, J. Carbonell, and E. Xing. Smoothing proximal gradient method for general structured sparse learning. *UAI*, 2011.

4. A. Dempster. Covariance selection. *Biometrics*, 1972.

5. J. Duchi, S. Gould, and D. Koller. Projected subgradient methods for learning sparse Gaussians. *UAI*, 2008.

6. J. Duchi and Y. Singer. Boosting with structural sparsity. *ICML*, 2009.

7. J. Duchi and Y. Singer. Efficient learning using forward-backward splitting. *NIPS*, 2009.

8. J. Friedman, T. Hastie, and R. Tibshirani. Sparse inverse covariance estimation with the graphical lasso. *Biostatistics*, 2007.

9. J. Friedman, T. Hastie, and R. Tibshirani. Applications of the lasso and grouped lasso to the estimation of sparse graphical models. *Technical Report, Stanford University*, 2010.

10. R. Goldstein, N. Alia-Klein, D. Tomasi, J. Honorio, T. Maloney, P. Woicik, R. Wang, F. Telang, and N. Volkow. Anterior cingulate cortex hypoactivations to an emotionally salient task in cocaine addiction. *Proceedings of the National Academy of Sciences, USA*, 2009.

11. R. Goldstein, D. Tomasi, N. Alia-Klein, L. Zhang, F. Telang, and N. Volkow. The effect of practice on a sustained attention task in cocaine abusers. *NeuroImage*, 2007.

12. J. Honorio, L. Ortiz, D. Samaras, N. Paragios, and R. Goldstein. Sparse and locally constant Gaussian graphical models. *NIPS*, 2009.

13. J. Honorio and D. Samaras. Multi-task learning of Gaussian graphical models. *ICML*, 2010.

14. J. Honorio, D. Samaras, I. Rish, and G. Cecchi. Variable selection for gaussian graphical models. In *International Conference on Artificial Intelligence and Statistics (AISTATS-12)*, pages 538–546, 2012.

15. S. Lauritzen. *Graphical Models*. Oxford Press, 1996.

16. J. Lee, F. Telang, C. Springer, and N. Volkow. Abnormal brain activation to visual stimulation in cocaine abusers. *Life Sciences*, 2003.

17. S. Lee, V. Ganapathi, and D. Koller. Efficient structure learning of Markov networks using ℓ_1-regularization. *NIPS*, 2006.

18. S. Lee, H. Lee, P. Abbeel, and A. Ng. Efficient ℓ_1 regularized logistic regression. *AAAI*, 2006.

19. E. Levina, A. Rothman, and J. Zhu. Sparse estimation of large covariance matrices via a nested lasso penalty. *The Annals of Applied Statistics*, 2008.

20. Q. Liu and A. Ihler. Learning scale free networks by reweighted ℓ_1 regularization. *AISTATS*, 2011.

21. J. Mairal, R. Jenatton, G. Obozinski, and F. Bach. Network flow algorithms for structured sparsity. *NIPS*, 2010.

22. B. Marlin and K. Murphy. Sparse Gaussian graphical models with unknown block structure. *ICML*, 2009.

23. B. Marlin, M. Schmidt, and K. Murphy. Group sparse priors for covariance estimation. *UAI*, 2009.

24. L. Meier, S. van de Geer, and P. Bühlmann. The group lasso for logistic regression. *Journal of the Royal Statistical Society*, 2008.

25. N. Meinshausen and P. Bühlmann. High dimensional graphs and variable selection with the lasso. *The Annals of Statistics*, 2006.

26. G. Natsoulis, L. El Ghaoui, G. Lanckriet, A. Tolley, F. Leroy, S. Dunlea, B. Eynon, C. Pearson, S. Tugendreich, and K. Jarnagin. Classification of a large microarray data set: algorithm comparison and analysis of drug signatures. *Genome Research*, 2005.

27. G. Obozinski, B. Taskar, and M. Jordan. Joint covariate selection and joint subspace selection for multiple classification problems. *Statistics and Computing*, 2010.

28. R. Parr, L. Li, G. Taylor, C. Painter-Wakefield, and M. Littman. An analysis of linear models, linear value-function approximation, and feature selection for reinforcement learning. *ICML*, 2008.

29. A. Quattoni, X. Carreras, M. Collins, and T. Darrell. An efficient projection for $\ell_{1,\infty}$ regularization. *ICML*, 2009.

30. M. Schmidt, K. Murphy, G. Fung, and R. Rosales. Structure learning in random fields for heart motion abnormality detection. *CVPR*, 2008.

31. M. Schmidt, A. Niculescu-Mizil, and K. Murphy. Learning graphical model structure using ℓ_1-regularization paths. *AAAI*, 2007.

32. M. Schmidt, E. van den Berg, M. Friedlander, and K. Murphy. Optimizing costly functions with simple constraints: A limited-memory projected quasi-Newton algorithm. *AISTATS*, 2009.

33. R. Tibshirani. Regression shrinkage and selection via the lasso. *Journal of the Royal Statistical Society*, 1996.

34. J. Tropp. Algorithms for simultaneous sparse approximation, part II: convex relaxation. *Signal Processing*, 2006.

35. B. Turlach, W. Venables, and S. Wright. Simultaneous variable selection. *Technometrics*, 2005.

36. M. Wainwright, P. Ravikumar, and J. Lafferty. High dimensional graphical model selection using ℓ_1-regularized logistic regression. *NIPS*, 2006.

37. M. Yuan and Y. Lin. Model selection and estimation in regression with grouped variables. *Journal of the Royal Statistical Society*, 2006.

38. M. Yuan and Y. Lin. Model selection and estimation in the Gaussian graphical model. *Biometrika*, 2007.

39. B. Zhang and Y. Wang. Learning structural changes of Gaussian graphical models in controlled experiments. *UAI*, 2010.

3   Log-Nonlinear Formulations for Robust High-dimensional Modeling

Aurélie C. Lozano   aclozano@us.ibm.com
IBM T. J. Watson Research Center
Yorktown Heights, NY, USA

Huijing Jiang   huijiang@us.ibm.com
IBM T. J. Watson Research Center
Yorktown Heights, NY, USA

Xinwei Deng   xdeng@vt.edu
Virginia Tech
Blacksburg, VA, USA

In this chapter we depart from log-linear models to consider "log-non-linear" formulations. Such formulations yield to non convex problems but provide significant advantage in terms of robustness. More specifically, we propose a robust framework to jointly perform two key modeling tasks involving high dimensional data: (i) learning a sparse functional mapping from multiple predictors to multiple responses while taking advantage of the coupling among responses, and (ii) estimating the conditional dependency structure among responses while adjusting for their predictors. The traditional likelihood-based estimators lack resilience with respect to outliers and model misspecification. This issue is exacerbated when dealing with high dimensional noisy data. In this work, we propose instead to minimize a regularized distance criterion, which is motivated by the minimum distance functionals used in nonparametric methods for their excellent robustness properties. The proposed estimates can be obtained efficiently by leveraging a sequential quadratic programming algorithm. We provide theoretical justification such as estimation consistency for the proposed estimator. Additionally, we shed light on the robustness of our estimator through its linearization, which yields a combination of weighted lasso and graphical lasso with the sample weights providing an intuitive explanation of the robustness. We demonstrate the merits of our framework through simulation study and the analysis of real financial and genetics data.

3.1 Introduction

This chapter departs from log-linear modeling to consider log-non-linear formulations for robust estimation from high-dimensional data. We focus on multiresponse regression where both predictor and response spaces may exhibit high dimensions. We propose a *robust* framework to jointly and synergistically solve two important tasks: (i) learning the sparse functional mapping between inputs and outputs while taking advantage of the coupling among responses, and (ii) estimating the conditional dependency structure among responses while adjusting for the covariates. This is motivated by the crucial need of integrating genomic and transcriptomic datasets in computational biology in order to solve two fundamental problems effectively: identifying the genetic variations in the genome that influence gene expression levels (a.k.a. expression quantitative trait loci eQTLs mapping), and uncovering gene expression networks. In fact, the accuracy of the first problem can then be improved by leveraging the gene relatedness, and similarly the accurate and faithful estimation of the gene expression networks can be obtained by accounting for the confounding genetic effects on gene expression.

Multiresponse regression (5) generalizes the basic single-response regression to model multiple responses that might significantly correlate with each other. As opposed to treating each response independently, one can jointly learn multiple regression mappings to improve the estimation and prediction accuracy by exploiting the conditional dependencies among responses. Variable selection in multiresponse regression can be accomplished via the penalized approaches including lasso (30) and multitask lasso (23).

Sparse estimation of inverse covariance matrix is an important area in the multivariate analysis with broad applications in graphical models. A major focus in this area is that of penalized maximum likelihood formulations (14, 33, 1, 15). In particular, Friedman et al. developed the graphical lasso (Glasso) (14), a block coordinate descent procedure to learn a sparse inverse covariance matrix. More recently, Hsieh et al. (15) proposed a quadratic approximation approach (QUIC) to solve the Glasso problem, which is extremely efficient especially when the inverse covariance is reasonably sparse.

Alternatively, modified Cholesky decompositions based on the likelihood can be used to estimate the sparse inverse covariance (16, 4, 20). A simpler approach of "neighborhood selection" (21) estimates sparse graphical models using lasso to regress on each variable with the others as predictors.

Combining multiresponse regression and inverse covariance estimation has recently begun to attract more attention in the machine learning community. Rothman et al. (26) proposed a multivariate regression with covariance estimation (MRCE) to jointly estimate the sparse regression and inverse covariance matrices. They demonstrated that exploiting the correlation structures can significantly improve the prediction accuracy. The same model was also studied by Lee and Liu (19) who provided some theoretical properties for their developed method. An alternative parameterization was considered by Sohn and Kim (28), which is based on the joint distribution of predictors and responses and yields an l_1-penalized conditional graphical model (l_1-CGGM). Another relevant method is that of covariate adjusted precision matrix estimation (CAPME) (7), a two-stage approach to estimate the conditional dependency structure among response variables by adjusting for covariates. The first stage is to estimate the regression coefficients via a multivariate extension of the l_1 Dantzig selector (8), and the second stage is to estimate the inverse covariance matrix using l_∞ error with an l_1 penalty.

Robustness is an important aspect often overlooked in the sparse learning literature, while critical when dealing with high dimensional noisy data. Traditional likelihood-based estimators such as MRCE and l_1-CGGM lack resilience to outliers and model misspecification. Additionally, to the best of our knowledge, estimates based on Dantzig selector have not been compared to lasso counterparts in terms of robustness. Thus it is unclear whether CAPME, for instance, can address the robustness issue. There is limited existing work on robust sparse learning methods in high-dimensional modeling. The LAD-lasso (31) performs single response regression using the least absolute deviation combined with an l_1 penalty. The tlasso (13) performs inverse covariance estimation using penalized log-likelihood with the multivariate t distribution. However, neither of these methods can be easily extended to the setting considered in this chapter.

We propose a robust approach to jointly estimate multiresponse regression and inverse covariance matrix. Our approach is based on a regularized distance criterion motivated by minimum distance estimators. Minimum distance estimators (32) are popularized in nonparametric methods and have exhibited excellent robustness properties (3, 11). Their use for parametric estimation has been discussed in (27, 2). In this work, we propose a penalized minimum distance criterion for robust estimation of sparse parametric models in the high dimensional settings. *Our key contributions* to this robust approach are as follows.

- The objective, which is denoted as REG-ISE, is based on the integrated squared error distance (ISE) between the model and the "true" distribution, and imposes the sparse model structure by adding sparsity-inducing penalties to the ISE criterion.

- Theoretical guarantees are provided on the estimation consistency of the proposed REG-ISE estimator.

- We leverage a sequential quadratic programming algorithm (9) to efficiently solve our objective.

- We shed light into the robustness of our framework by linearizing our objective. The linearization yields a problem combining weighted versions of l_1-penalized regression (lasso) and l_1-penalized inverse covariance estimation (glasso), where the weights assigned to the instances are theoretically derived and can be interpreted in terms of "outlying degrees".

The strength of our method is demonstrated via simulation data with and without outliers. Our study also confirms that outliers can severely influence the variable selection accuracy of some existing sparse learning methods. Experiments on real financial and eQTL data further illustrate the merits of the proposed method.

The remainder of this chapter is organized as follows. In Section 3.2 and Section 3.3 we present our robust joint estimation methodology, its optimization algorithm, and a modified cross-validation method for tuning parameter selection. We perform extensive simulation in Section 3.4 to demonstrate the performance of our method compared to MRCE, LAD-lasso, and CAPME for settings with or without outliers. In Section 3.5.2 we apply our method to the analysis of real financial and biological datasets. Section 3.6 concludes our paper.

3.2   Model Setup

Denote the response vector $\boldsymbol{y} = (y_1, \ldots, y_q)' \in \mathcal{R}^q$ and the predictor vector $\boldsymbol{x} = (x_1, \ldots, x_p)' \in \mathcal{R}^p$. We consider a multiresponse linear regression model

$$\boldsymbol{y} = \boldsymbol{B}'\boldsymbol{x} + \boldsymbol{\epsilon}, \quad \boldsymbol{\epsilon} \sim \mathcal{N}(0, \boldsymbol{\Sigma}), \tag{3.1}$$

where $\boldsymbol{B} = (b_{ij})$ is a $p \times q$ matrix of coefficients and the kth column is the coefficients associated with kth response y_k regressing on the predictors $\boldsymbol{x}$. The $q \times q$ covariance matrix $\boldsymbol{\Sigma}$ describes the covariance structure of response vector $\boldsymbol{y}$ given the predictors $\boldsymbol{x}$. Moreover, its inverse $\boldsymbol{\Sigma}^{-1} = (c_{ij})$ represents the partial covariance structure (18) and has been widely used to learn a sparse graphical model under Gaussian assumption. Note that $\mathbf{\Sigma}^{-1}$ in (3.1) captures the partial covariances among responses $\mathbf{y}$ after adjusting for the effects of covariates $\mathbf{x}$. For simplicity of notation, we assume the data are centered so that the model (3.1) does not contain intercepts.

Suppose there are n observational vectors $\mathbf{x}_i = (x_{i1}, \ldots, x_{ip})'$, $i = 1, \ldots, n$ and the corresponding response vectors are $\mathbf{y}_i = (y_{i1}, \ldots, y_{iq})'$. To jointly obtain sparse estimates of coefficient matrix $\mathbf{B}$ and precision matrix $\mathbf{\Sigma}^{-1}$, we consider a loss function $L_n(\mathbf{B}, \mathbf{\Sigma})$ that measures the goodness-of-fit on the multivariate response. The sparse structures of $\mathbf{B}$ and $\mathbf{\Sigma}^{-1}$ are encouraged by using l_1 penalties. Specifically, the penalized loss function $L_{n,\lambda}(\mathbf{B}, \mathbf{\Sigma})$ is written as

$$L_{n,\lambda}(\mathbf{B}, \mathbf{\Sigma}) = L_n(\mathbf{B}, \mathbf{\Sigma}) + \lambda_1 \|\mathbf{\Sigma}^{-1}\|_1 + \lambda_2 \|\mathbf{B}\|_1. \tag{3.2}$$

where $\|\mathbf{\Sigma}^{-1}\|_1 = \sum_{i \leq j} |c_{ij}|$ and $\|\mathbf{B}\|_1 = \sum_{i,j} |b_{ij}|$ are l_1 matrix norms. Following the principle of parsimony, we consider l_1 penalty functions to seek a most appropriate model that adequately explains the data. With carefully selected tuning parameters λ_1 and λ_2, we can achieve an optimal trade-off between the parsimoniousness and goodness-of-fit of the model. We note that general structured penalties can be used in lieu of the l_1 norm, such as multi-task lasso penalties (23).

The loss $L_n(\mathbf{B}, \mathbf{\Sigma})$ is typically derived from a likelihood-based approach. For instance the MRCE method (26) uses

$$L_n(\mathbf{B}, \mathbf{\Sigma}) = \text{trace}\left((\mathbf{Y} - \mathbf{X}\mathbf{B})^T(\mathbf{Y} - \mathbf{X}\mathbf{B})\mathbf{\Sigma}^{-1}\right) - \log|\mathbf{\Sigma}^{-1}|.$$

Altenatively, if one ignores the contribution of the inverse covariance matrix (by implicitely assuming that it is the identity matrix), one can consider the traditional squared loss $L_n = \|\mathbf{Y} - \mathbf{X}\mathbf{B}\|_F^2$ and a straighforward generalization of the traditional Lasso estimator (30) to the mutiresponse setting.

3.3 A Regularized Integrated Squared Error Estimator

We begin this section by showing how a minimum distance criterion yields our proposed estimator for achieving robustness under the model in (3.1).

3.3.1   Derivation of the REG-ISE Objective

We first apply the Integrated Squared Error (ISE) criterion to the conditional distribution of response vector $\boldsymbol{y}$ given the predictors $\boldsymbol{x}$. It leads to an L_2 distance between the true conditional distribution $f(\boldsymbol{y}|\boldsymbol{x})$ and the parametric distribution function $f(\boldsymbol{y}|\boldsymbol{x}; \boldsymbol{B}, \boldsymbol{\Sigma})$ as follows

$$
\begin{aligned}
\tilde{L}(\boldsymbol{B}, \boldsymbol{\Sigma}) &= \int \left[f(\boldsymbol{y}|\boldsymbol{x}; \boldsymbol{B}, \boldsymbol{\Sigma}) - f(\boldsymbol{y}|\boldsymbol{x}) \right]^2 d\boldsymbol{y} \\
&= \int f^2(\boldsymbol{y}|\boldsymbol{x}; \boldsymbol{B}, \boldsymbol{\Sigma}) d\boldsymbol{y} - 2 \int f(\boldsymbol{y}|\boldsymbol{x}; \boldsymbol{B}, \boldsymbol{\Sigma}) f(\boldsymbol{y}|\boldsymbol{x}) d\boldsymbol{y} \\
&\quad + \int f^2(\boldsymbol{y}|\boldsymbol{x}) d\boldsymbol{y} \\
&= \int f^2(\boldsymbol{y}|\boldsymbol{x}; \boldsymbol{B}, \boldsymbol{\Sigma}) d\boldsymbol{y} - 2\mathbb{E}[f(\boldsymbol{y}|\boldsymbol{x}; \boldsymbol{B}, \boldsymbol{\Sigma})] + \text{constant},
\end{aligned}
$$

where $f(\boldsymbol{y}|\boldsymbol{x}; \boldsymbol{B}, \boldsymbol{\Sigma})$ is the probability density function of multivariate normal $\mathcal{N}(\boldsymbol{B}'\boldsymbol{x}, \boldsymbol{\Sigma})$ and $\int f(\boldsymbol{y}|\boldsymbol{x})^2 d\boldsymbol{y}$ is a constant independent of $\boldsymbol{B}$ and $\boldsymbol{\Sigma}$.

We have that

$$
f(\boldsymbol{y}|\boldsymbol{x}; \boldsymbol{B}, \boldsymbol{\Sigma}) \equiv f(\boldsymbol{y} - \boldsymbol{B}'\boldsymbol{x}; 0, \boldsymbol{\Sigma})
$$

because of the conditional distribution assumption. Since $\boldsymbol{\epsilon} = \boldsymbol{y} - \boldsymbol{B}'\boldsymbol{x}$ are independently and identically distributed, one can consider approximating $\mathbb{E}[f(\boldsymbol{y}|\boldsymbol{x}; \boldsymbol{B}, \boldsymbol{\Sigma})]$ by the empirical mean

$$
\frac{1}{n} \sum_{i=1}^{n} f(\boldsymbol{y}_i|\boldsymbol{x}_i; \boldsymbol{B}, \boldsymbol{\Sigma}).
$$

Similar approximation techniques have also been used for Gaussian mixture density estimation (27). Now the resulting empirical loss function of $\tilde{L}(\boldsymbol{B}, \boldsymbol{\Sigma})$ can therefore be written as

$$
\tilde{L}_n(\boldsymbol{B}, \boldsymbol{\Sigma}) = \int f^2(\boldsymbol{y}|\boldsymbol{x}; \boldsymbol{B}, \boldsymbol{\Sigma}) d\boldsymbol{y} - \frac{2}{n} \sum_{i=1}^{n} f(\boldsymbol{y}_i|\boldsymbol{x}_i; \boldsymbol{B}, \boldsymbol{\Sigma}) \tag{3.3}
$$

where

$$
\int f^2(\boldsymbol{y}|\boldsymbol{x}; \boldsymbol{B}, \boldsymbol{\Sigma}) d\boldsymbol{y} = 1/(2^q \pi^{q/2} |\boldsymbol{\Sigma}|^{1/2}).
$$

Note that we assume a parametric family for the model while using a nonparametric ISE criterion to measure goodness of fit.

From the perspective of the loss function, ISE is a more robust measure of the goodness-of-fit compared with the likelihood-based loss function. It can match the model with the largest portion of the data because the integration in (3.3) accounts for the whole range of the squared loss function.

Using ISE criterion as the loss function, the objective function in (3.2) becomes

$$\tilde{L}_{n,\lambda}(\boldsymbol{B}, \boldsymbol{\Sigma}) = \frac{|\boldsymbol{\Sigma}^{-1}|^{1/2}}{2^q \pi^{q/2}} - \frac{2}{n} \sum_{i=1}^{n} f(\boldsymbol{y}_i | \boldsymbol{x}_i; \boldsymbol{B}, \boldsymbol{\Sigma})$$
$$+ \lambda_1 \|\boldsymbol{\Sigma}^{-1}\|_1 + \lambda_2 \|\boldsymbol{B}\|_1. \tag{3.4}$$

However, the minimization of the objective in (3.4) is challenging. To circumvent this difficulty, we consider minimizing an upper bound of (3.4) which retains the robustness property.

For that purpose, we introduce a lemma which is essential for deriving the proposed objective (3.8) for estimating the sparse multiresponse regression model in (3.1).

Lemma 3.1. *For a positive definite matrix $\boldsymbol{\Sigma}^{-1}$ with dimension q, the relation between its determinant value and l_1 norm can be described in the following inequality $|\boldsymbol{\Sigma}^{-1}|^{1/2} \leq \left(\frac{\|\boldsymbol{\Sigma}^{-1}\|_1}{q} \right)^{q/2}$.*

Proof. Suppose that the eigenvalues of $\boldsymbol{\Sigma}^{-1}$ are $d_i, i = 1, \ldots, q$. Using the fact that $\sqrt[q]{\prod_{i=1}^{q} d_i} \leq \frac{1}{q} \sum_{i=1}^{q} d_i$, one can have

$$|\boldsymbol{\Sigma}^{-1}|^{1/2} \leq \left(\frac{\sum_{i=1}^{q} d_i}{q} \right)^{q/2}.$$

For each eigenvalue d_i, we apply the Gershgorin's circle theorem to obtain its upper bound. That is

$$|d_i - c_{ii}| \leq \sum_{j \neq i} |c_{ij}| \Rightarrow d_i \leq \sum_{j=1}^{q} |c_{ij}|$$

where $c_{ij}, i = 1, \ldots, q; j = 1, \ldots, q$ are the elements in matrix $\boldsymbol{\Sigma}^{-1}$. Therefore, we have $\sum_{i=1}^{q} d_i \leq \sum_{i=1}^{q} \sum_{j=1}^{q} |c_{ij}| = \|\boldsymbol{\Sigma}^{-1}\|_1$, which leads to

$$|\boldsymbol{\Sigma}^{-1}|^{1/2} \leq \left(\frac{\|\boldsymbol{\Sigma}^{-1}\|_1}{q} \right)^{q/2}.$$

This ends the proof of Lemma 1. □

Using Lemma 3.1, we can derive an upper bound for the objective function (3.4) as follows

$$c^* \|\boldsymbol{\Sigma}^{-1}\|_1^{q/2} - \frac{2}{n} \sum_{i=1}^{n} f(\boldsymbol{y}_i | \boldsymbol{x}_i; \boldsymbol{B}, \boldsymbol{\Sigma}) + \lambda_1 \|\boldsymbol{\Sigma}^{-1}\|_1 + \lambda_2 \|\boldsymbol{B}\|_1, \tag{3.5}$$

where $c^* = 2^{-q}(\pi q)^{-q/2}$ is a constant. The above optimization problem amounts to minimizing

$$\breve{L}_{n,\lambda}(\boldsymbol{B}, \boldsymbol{\Sigma}) = \breve{L}_n(\boldsymbol{B}, \boldsymbol{\Sigma}) + \lambda_1^* \|\boldsymbol{\Sigma}^{-1}\|_1 + \lambda_2 \|\boldsymbol{B}\|_1, \tag{3.6}$$

where $\breve{L}_n(\boldsymbol{B}, \boldsymbol{\Sigma}) = -\frac{2}{n}\sum_{i=1}^{n} f(\boldsymbol{y}_i|\boldsymbol{x}_i; \boldsymbol{B}, \boldsymbol{\Sigma})$ and λ_1^* is appropriately chosen. The value of λ_1^* here should be slightly larger than the value of λ_1 in (3.4). Moreover, the diagonal elements of $\boldsymbol{\Sigma}^{-1}$ are also penalized.

Note that $\breve{L}_{n,\lambda}(\boldsymbol{B}, \boldsymbol{\Sigma})$ is an upper bound of $\tilde{L}_{n,\lambda}(\boldsymbol{B}, \boldsymbol{\Sigma})$, however, the difference $\tilde{L}_{n,\lambda}(\boldsymbol{B}, \boldsymbol{\Sigma}) - \breve{L}_{n,\lambda}(\boldsymbol{B}, \boldsymbol{\Sigma})$ is well controlled by the penalty term $\lambda_1^* \|\boldsymbol{\Sigma}^{-1}\|_1$ in (3.6). By properly adjusting the value of λ_1^*, we can make the difference reasonably small. Therefore, by minimizing $\breve{L}_{n,\lambda}(\boldsymbol{B}, \boldsymbol{\Sigma})$, we expect to approach the solution of $\tilde{L}_{n,\lambda}(\boldsymbol{B}, \boldsymbol{\Sigma})$ and thus still retain the robustness property in the estimators.

Taking the logarithm on $\breve{L}_n(\boldsymbol{B}, \boldsymbol{\Sigma})$, we obtain the loss

$$L_n(\boldsymbol{B}, \boldsymbol{\Sigma}) = -\log\left[\frac{1}{n}\sum_{i=1}^{n}\exp(-\frac{1}{2}(\boldsymbol{y}_i - \boldsymbol{B}'\boldsymbol{x}_i)'\boldsymbol{\Sigma}^{-1}(\boldsymbol{y}_i - \boldsymbol{B}'\boldsymbol{x}_i))\right]$$
$$-\frac{1}{2}\log|\boldsymbol{\Sigma}^{-1}|. \tag{3.7}$$

We note that the logarithm is employed to strike a better balance between goodness of fit and the sparsity inducing penalty (similarly one considers the penalized negative log-likelihood rather than dealing with the likelihood directly). This yields the estimator proposed and studied in this chapter, the Regularized Integrated Square Error (REG-ISE) estimator, which minimizes the following objective function:

$$L_{n,\lambda}(\boldsymbol{B}, \boldsymbol{\Sigma}) =$$
$$-\log\left[\frac{1}{n}\sum_{i=1}^{n}\exp(-\frac{1}{2}(\boldsymbol{y}_i - \boldsymbol{B}'\boldsymbol{x}_i)'\boldsymbol{\Sigma}^{-1}(\boldsymbol{y}_i - \boldsymbol{B}'\boldsymbol{x}_i))\right]$$
$$-\frac{1}{2}\log|\boldsymbol{\Sigma}^{-1}| + \lambda_1\|\boldsymbol{\Sigma}^{-1}\|_1 + \lambda_2\|\boldsymbol{B}\|_1. \tag{3.8}$$

For notational convenience, here and in later sections, we use λ_1 for λ_1^*.

Some intuition can already be gained on the objective robustness by considering the ratio between data and model pdf: $f(\boldsymbol{y}|\boldsymbol{x})/f(\boldsymbol{y}|\boldsymbol{x}; \boldsymbol{B}, \boldsymbol{\Sigma})$. An outlier in the data may drives this ratio to infinity, in which case the log-likelihood is also infinity. In contrast, the difference $f(\boldsymbol{y}|\boldsymbol{x}) - f(\boldsymbol{y}|\boldsymbol{x}; \boldsymbol{B}, \boldsymbol{\Sigma})$ is always bounded. This property makes the L_2-distance a favourable choice when dealing with outliers. We note that a similar reasoning holds in the context of density estimation, as pointed out in the recent work of (29) on density-difference estimation.

3.3.2 Optimization

The REG-ISE objective function (3.8) is non-convex and non-smooth. In order to solve it, one could consider approximating the "log-sum-exp" term in the objective, combined with alternate optimization for B and Σ^{-1} respectively (as done in MRCE). However, the convergence of alternate optimization can be very slow, as observed in the case of MRCE (26).

Instead, we propose to adopt a sequential quadratic programming algorithm recently developed by Curtis & Overton (9) for the non-smooth and non-convex optimization, named SLQP-GS. The basic idea of SLQP-GS is to combine sequential quadratic approximation with a process of gradient sampling so that the computation of the search direction is effective in non-smooth regions. The only requirement for SLQP-GS to be applicable is that the objective and constraints (if any) be continuously differentiable on open dense subsets, which is satisfied in our case. We also benefit from the convergence guarantees of SLQP-GS, namely that the algorithm is guaranteed to converge to a solution regardless of the initialization with probability one.

We employed the Matlab implementation of SLQP-GS provided by the authors, which is available from `http://coral.ie.lehigh.edu/~frankecurtis/software`. We refer the reader to Curtis & Overton (9) for details on the algorithm, its matlab implementation and convergence results. We note that the gradient sampling step in SLQP-GS can be efficiently parallelized for fast computation in high dimensional applications. Alternatively one can perform adaptive sampling of gradients over the course of the optimization process as described in (10).

3.3.3 Consistency Results

The L_2 distance estimators are known to strike the right balance between statistical efficiency and robustness (2, 27). In this section, we add to this body of evidence by showing that the REG-ISE estimator is root-n consistent for the settings of the fixed dimensionality p. Denote by $\bar{B}$ the true regression coefficient matrix, and by $\bar{\Sigma}$ the true covariance matrix. We assume the following conditions:

(C1) $\frac{1}{n}X'X \to A$, where A is positive definite.

(C2) There exist $\sqrt{n}$-consistent estimators of $\bar{B}$ and $\bar{\Sigma}^{-1}$.

Condition (C2) can be replaced by some technical regularity conditions as in (12) so as to guarantee the consistency of ordinary maximum likelihood estimators.

Theorem 3.2. *Consider sequences $\lambda_{1,n}$ and $\lambda_{2,n}$ of regularization parameters, such that*

$$\lambda_{1,n}n^{-1/2} \to 0 \text{ and } \lambda_{2,n}n^{-1/2} \to 0.$$

Then under the conditions (C1) and (C2), there exists a local minimizer $(\hat{\boldsymbol{B}}, \hat{\boldsymbol{\Sigma}}^{-1})$ of REG-ISE such that

$$\|(vec(\hat{\boldsymbol{B}})', vec(\hat{\boldsymbol{\Sigma}}^{-1})')' - (vec(\bar{\boldsymbol{B}})', vec(\bar{\boldsymbol{\Sigma}}^{-1})')\|_2 = O_p(1/\sqrt{n}).$$

A proof is provided in the Appendix.

The above theoretical results hold for the case where the dimensionality p is fixed, while the sample size n is allowed to grow. As a future work we plan to extend our results to the case where p is allowed to grow with the sample size n. We conjecture that condition (C1) might be replaced by conditions on sample and population covariance while (C2) is still achieved by certain penalized maximum likelihood estimators (22).

3.3.4 Insights into Robustness

We provide some insights for the robustness of REG-ISE by considering a first order approximation of the "log-sum-exp" term in (3.8).

Define the parameter set $\boldsymbol{\beta} = (\boldsymbol{B}, \boldsymbol{\Sigma}^{-1})$ and denote

$$g_i(\boldsymbol{\beta}) = -\frac{1}{2}(\boldsymbol{y}_i - \boldsymbol{B}'\boldsymbol{x}_i)'\boldsymbol{\Sigma}^{-1}(\boldsymbol{y}_i - \boldsymbol{B}'\boldsymbol{x}_i).$$

We consider a first-order approximation for $\log\left[\frac{1}{n}\sum_{i=1}^n \exp(g_i(\boldsymbol{\beta}))\right]$ with respect to $\boldsymbol{\beta}$ as follows:

$$\log\left[\frac{1}{n}\sum_{i=1}^n \exp(g_i(\boldsymbol{\beta}))\right] \approx C_0 + \frac{1}{n}\sum_{i=1}^n \frac{\exp(g_i(\boldsymbol{\beta}_0))}{\frac{1}{n}\sum_{i=1}^n \exp(g_i(\boldsymbol{\beta}_0))}\nabla g_i(\boldsymbol{\beta}_0)^T(\boldsymbol{\beta} - \boldsymbol{\beta}_0),$$

where $\boldsymbol{\beta}_0$ is an initial estimate and C_0 is some constant independent of $\boldsymbol{\beta}$.

Using the fact that $g_i(\boldsymbol{\beta}) \approx g_i(\boldsymbol{\beta}_0) + \nabla g_i(\boldsymbol{\beta}_0)^T(\boldsymbol{\beta} - \boldsymbol{\beta}_0)$, we have the following

$$\log\left[\frac{1}{n}\sum_{i=1}^n \exp(g_i(\boldsymbol{\beta}))\right] \propto \frac{1}{n}\sum_{i=1}^n \frac{\exp(g_i(\boldsymbol{\beta}_0))}{\frac{1}{n}\sum_{i=1}^n \exp(g_i(\boldsymbol{\beta}_0))}g_i(\boldsymbol{\beta}),$$

up to some constant independent of $\boldsymbol{\beta}$. Therefore, the objection function

(3.8) can be approximated by

$$- \log |\mathbf{\Sigma}^{-1}| + \frac{1}{n} \sum_{i=1}^{n} w_i (\mathbf{y}_i - \mathbf{B}' \mathbf{x}_i)' \mathbf{\Sigma}^{-1} (\mathbf{y}_i - \mathbf{B}' \mathbf{x}_i)$$

$$+ \lambda_1 \|\mathbf{\Sigma}^{-1}\|_1 + \lambda_2 \|\mathbf{B}\|_1, \tag{3.9}$$

up to some constant and where

$$w_i \equiv w_i(\boldsymbol{\beta}_0) = \frac{\exp(g_i(\boldsymbol{\beta}_0))}{\frac{1}{n} \sum_{i=1}^{n} \exp(g_i(\boldsymbol{\beta}_0))}. \tag{3.10}$$

By defining

$$\mathbf{S}^* = \mathbf{S}^*(\boldsymbol{\beta}_0) = \frac{1}{n} \sum_{i=1}^{n} w_i (\mathbf{y}_i - \mathbf{B}' \mathbf{x}_i)(\mathbf{y}_i - \mathbf{B}' \mathbf{x}_i)',$$

we can rewrite (3.9) as

$$- \log |\mathbf{\Sigma}^{-1}| + \text{trace}[\mathbf{\Sigma}^{-1} \mathbf{S}^*(\boldsymbol{\beta}_0)] + \lambda_1 \|\mathbf{\Sigma}^{-1}\|_1 + \lambda_2 \|\mathbf{B}\|_1. \tag{3.11}$$

Note that $\mathbf{S}^*$ can be viewed as a weighted sample covariance matrix where weights are with respect to n observations.

One can then envision an approximate iterative procedure where given initial estimates, data are first re-weighted by w_i in (3.10) and then alternately passed to lasso and l_1 penalized inverse covariance solvers (e.g. QUIC, glasso) to provide new estimates, and the procedure would be repeated until convergence (see details in the appendix). This intuitively elaborates the robustness property of REG-ISE. Indeed the weights w_i are proportional to the likelihood functions of individual data points, i.e., $w_i = \dfrac{L(\mathbf{y}_i | \mathbf{x}_i; \beta_0)}{\sum_{i=1}^{n} L(\mathbf{y}_i | \mathbf{x}_i; \beta_0)}$. Thus data with high likelihood values are given more weights in the estimation. Conversely, data with low likelihood values, which are more likely to be outliers, contribute less to the estimation. The connection between the likelihood functions and weights nicely explains the resilience of the proposed estimator to outliers. For completeness, the details of the procedure are presented in algorithm 3.1 The initial estimates $\mathbf{B}_0$ are obtained using ridge regression(26), followed by a well-conditioned estimate $\mathbf{\Sigma}_0^{-1}$ from the inverse of the sample covariance matrix of ridge regression residuals perturbed by a positive diagonal matrix

3.3.5 Tuning Parameter Selection

Approaches for choosing tuning parameters include cross-validation (CV) (4), the hold-out validation set method (20), and information criteria such as Bayesian information criterion (BIC) (33). Here we proposed a modified

---

Algorithm 3.1 Approximate Iterative Procedure

Step 1: Given an initial estimate $\boldsymbol{\Sigma}_0^{-1}$ and an initial estimate $\boldsymbol{B}_0$.

Step 2: Compute w_i based on (3.10) and obtain $\boldsymbol{S}^* = \frac{1}{n}\sum_{i=1}^n w_i(\boldsymbol{y}_i - \boldsymbol{B}_0'\boldsymbol{x}_i)(\boldsymbol{y}_i - \boldsymbol{B}_0'\boldsymbol{x}_i)'$.

Step 3: Estimate $\boldsymbol{\Sigma}^{-1}$ by minimizing (3.11) given $\boldsymbol{B}_0$:

$$\hat{\boldsymbol{\Sigma}}^{-1} = \arg\min -\log|\boldsymbol{\Sigma}^{-1}| + \text{trace}[\boldsymbol{\Sigma}^{-1}\boldsymbol{S}^*] + \lambda_1\|\boldsymbol{\Sigma}^{-1}\|_1.$$

Step 4: Estimate $\boldsymbol{B}$ by minimizing (3.9) given $\boldsymbol{\Sigma}_0^{-1}$:

$$\hat{\boldsymbol{B}} = \arg\min \frac{1}{n}\sum_{i=1}^n w_i(\boldsymbol{y}_i - \boldsymbol{B}'\boldsymbol{x}_i)'\boldsymbol{\Sigma}_0^{-1}(\boldsymbol{y}_i - \boldsymbol{B}'\boldsymbol{x}_i) + \lambda_2\|\boldsymbol{B}\|_1.$$

Step 5: If $\|\hat{\boldsymbol{\Sigma}}^{-1} - \boldsymbol{\Sigma}_0^{-1}\|_F^2 \le \delta_1$ and $\|\hat{\boldsymbol{B}} - \boldsymbol{B}_0\|_F^2 \le \delta_2$, stop. Else set $\boldsymbol{\Sigma}_0^{-1} = \hat{\boldsymbol{\Sigma}}^{-1}$ and $\boldsymbol{B}_0 = \hat{\boldsymbol{B}}$ and go back to Step 2.

---

scheme for the cross-validation method. The common K-fold CV consists in randomly partitioning the data into K folds, and then leaving out one fold of data as validation set while all the other folds are used as training set in each CV iteration. Note that CV assumes that the data are i.i.d. distributed, and therefore the validation set and training set are considered statistically equivalent. However, such an assumption is no longer valid in the presence of outliers since the proportions of the outliers in the validation data and training data can be different. Consequently, the validation set cannot be used to evaluate the model obtained by the training set.

To tackle this issue, we develop a modified cross-validation scheme motivated by the idea of sliced designs (24). Specifically, we perform K-fold cross-validation for $n = mK$ observations as follows. Based on initial estimates of the model parameters, we first rank the observed data according to the values of their likelihood functions. Then the first K data points are randomly assigned to K folds, one point per fold. Subsequently the next K data points are randomly assigned to K folds. This procedure is repeated m times. In this way, the data in each fold are more likely to have similar distributions. This modified scheme can also be applied to tuning via hold-out validation set method.

3.4   Simulation Study

We compare the proposed REG-ISE with MRCE (26), l_1-CGGM (28), CAPME (7), and LAD-Lasso (31). Note that LAD-Lasso can only estimate regression coefficients.

In our experiments, the rows of $n \times p$ predictor matrices $\boldsymbol{X}$ are sampled independently from $\mathcal{N}(\boldsymbol{0}, \boldsymbol{\Sigma_x})$ where $(\boldsymbol{\Sigma_x})_{i,j} = 0.5^{|i-j|}$. We consider the following two cases.

Case 1: The covariance matrix is set to $\Sigma_{i,j} = 0.7^{|i-j|}$ which corresponds to an AR(1) model with banded Σ^{-1}. We randomly select 10 percent of the predictors to be irrelevant to all the responses. Then for each response, we randomly select half of the remaining predictors to be relevant for that response. The corresponding non-zero entries in the regression matrix $\boldsymbol{B}$ are sampled independently from $\mathcal{N}(0,1)$. We consider 60 predictors, 20 responses and 100 observations.

Case 2: Σ^{-1} is the graph Laplacian of a tree with outdegree of 4 and edge weights uniformly sampled from $[0.3, 1.0]$. For each response we randomly select 10 percent of the predictors to be relevant, and sample the corresponding non-zero enties in $\boldsymbol{B}$ independently from $\mathcal{N}(0,1)$. We consider 1000 predictors, 100 responses and 400 observations.

To address the robustness issue, we consider various percentages of outliers contaminating the responses. The uncontaminated data are generated from $\boldsymbol{y} \sim \mathcal{N}(\boldsymbol{B}'\boldsymbol{x}, \boldsymbol{\Sigma})$, where $\boldsymbol{B}$ and $\boldsymbol{\Sigma}$ are specified above. Two scenarios presenting outliers are considered: (i) outliers with respect to the mean: $\boldsymbol{y} \sim \mathcal{N}(\boldsymbol{B}'\boldsymbol{x} + \boldsymbol{C}, \boldsymbol{\Sigma})$ where $\boldsymbol{C}$ is a constant vector of 5 and (ii) outliers regarding the covariance structure: $\boldsymbol{y} \sim \mathcal{N}(\boldsymbol{B}'\boldsymbol{x}, \boldsymbol{I})$ where $\boldsymbol{I}$ is an identity matrix.

To measure variable selection accuracy, we use the F_1 score defined by $F_1 = 2PR/(P+R)$, where P is precision (fraction of correctly selected variables among selected variables) and R is recall (fraction of correctly selected variables among true relevant variables). To measure the estimation accuracy of $\boldsymbol{B}$, we report the model error defined as $ME(\hat{\boldsymbol{B}}, \boldsymbol{B}) = \text{tr}\left[(\hat{\boldsymbol{B}} - \boldsymbol{B})^T \boldsymbol{\Sigma}_{\boldsymbol{x}} (\hat{\boldsymbol{B}} - \boldsymbol{B})\right]$, where $\hat{\boldsymbol{B}}$ is the estimated regression coefficient matrix. The estimation accuracy for $\boldsymbol{\Sigma}^{-1}$ is measured by its l_2 loss, defined as $\|\hat{\boldsymbol{\Sigma}}^{-1} - \boldsymbol{\Sigma}^{-1}\|_F$, under Frobenius norm where $\hat{\boldsymbol{\Sigma}}^{-1}$ is the estimated inverse covariance matrix.

For each of the above settings, we generate 50 simulated dataset. For all five comparison methods, for each dataset we use the modified 5-fold cross-validation described in Section 3.5 to tune parameters λ_1 and λ_2.

The choice of initial parameter estimates is important for MRCE and REG-ISE as their respective objective functions are non-convex. The initial estimates for $\boldsymbol{B}$ are obtained using ridge regression. In addition for REG-ISE, $\boldsymbol{\Sigma}$ is initialized as the inverse of the sample covariance matrix of ridge regression residuals perturbed by a positive diagonal matrix.

Figures 3.1 and 3.2 present the results for Case 1. Results for Case 2 are summarized in Table 3.1 Similar behavior in terms of robustness is observed for both cases. Our proposed REG-ISE estimator clearly outperforms other methods due to its robustness against outliers with respect to both mean and covariance deviations. Performance of MRCE, l_1-CGGM and CAPME seriously degrade once outliers are introduced. Surprisingly, LAD-Lasso does not show much resilience to outliers. Moreover, with "clean" data, its estimation and variable selection accuracy is inferior to other methods as it ignores the dependencies among responses. Interestingly, even when there are no outliers in the data, REG-ISE is competitive, since it is more likely to distinguish true signals from various noise amplitudes.

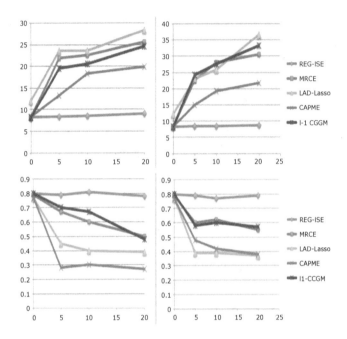

Figure 3.1: Average model error $ME(\hat{\boldsymbol{B}}, \boldsymbol{B})$ (top) and F_1 scores (bottom) for $\boldsymbol{B}$ estimated by REG-ISE, MRCE, l_1-CGGM, LAD-Lasso, and CAPME on simulated data of Case 1. Outliers in terms of the mean (left), and covariance (right). The x-axis corresponds to the percentage of outliers.

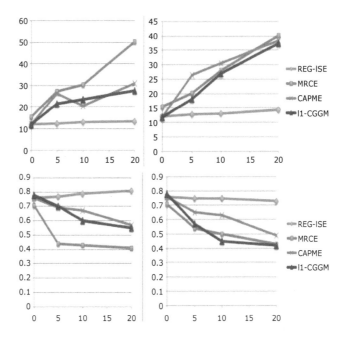

Figure 3.2: Average estimation error $\|\hat{\mathbf{\Sigma}}^{-1} - \mathbf{\Sigma}^{-1}\|_F$ (top) and F_1 scores (bottom) for $\mathbf{\Sigma}^{-1}$ estimated by REG-ISE, MRCE, l_1–CGGM, LAD-Lasso, and CAPME on simulated data of Case 1. Outliers in terms of the mean (left), and covariance (right). The x-axis corresponds to the percentage of outliers.

3.5 Applications

In this section, we illustrate the usefulness of the proposed robust methods through two motivating applications and compare the results of our robust estimators with those of existing methods.

3.5.1 Asset Return Prediction

As a toy example for multivariate time series, we analyze a financial dataset which has been studied in (26) and (33). This dataset contains weekly log-returns of 9 stocks for year 2004. Given multivariate time series data of log-returns $\boldsymbol{y}_t$ for weeks $t = 1, ..., T$, a first-order vector autoregressive model is considered as follows

$$\boldsymbol{y}_t = \boldsymbol{B}'\boldsymbol{y}_{t-1} + \boldsymbol{\epsilon}_t, \boldsymbol{\epsilon}_t \sim \mathcal{N}(\mathbf{0}, \mathbf{\Sigma}), t = 2, \ldots, T$$

where the response matrix $\boldsymbol{y}_t$ is formed by observations at week t and the predictor matrix $\boldsymbol{y}_{t-1}$ contains observations at the previous week $t - 1$. Following the analysis in Rothman et al. (26), we used log-returns of the

Table 3.1: Simulation results for Case 2. Top: Model error for B/l_2 loss for Σ^{-1}. Bottom: F_1 for B/F_1 for Σ^{-1}.

Measure	Outlier Type	Outlier %	REG-ISE	MRCE	1-1 CGGM	CAPME	LAD-Lasso
$ME(B)/l_2(\Sigma^{-1})$	None	0	33.12/48.3	33.92/61.88	32/47.6	33.4/49.6	47.96/NA
$ME(B)/l_2(\Sigma^{-1})$	Mean	5	33.44/49.28	87.56/109.4	78/75.8	53.12/59.3	94.04/NA
$ME(B)/l_2(\Sigma^{-1})$	Mean	10	34/52.84	90.28 / 178.2	82.4/80	73.2/72.79	94.24/NA
$ME(B)/l_2(\Sigma^{-1})$	Mean	20	35.96/53.4	102.6/196.3	98.4/101	79.36/80.23	113/NA
$ME(B)/l_2(\Sigma^{-1})$	Cov	5	34.04/51.04	90.96/81.04	97.36/62.8	60.2/96.34	92.68/NA
$ME(B)/l_2(\Sigma^{-1})$	Cov	10	34.29/52.6	96.39/102.1	110.16/104.2	77.28/156.2	103.6/NA
$ME(B)/l_2(\Sigma^{-1})$	Cov	20	36.5/56.92	122.6/186.4	133/167.1	87.4/153.4	146.68/NA
$F_1(B)/F_1(\Sigma^{-1})$	None	0	0.8/0.77	0.71/0.32	0.79/0.78	0.72/0.76	0.6/NA
$F_1(B)/F_1(\Sigma^{-1})$	Mean	5	0.78/0.79	0.65/0.24	0.65/0.7	0.28/0.59	0.45/NA
$F_1(B)/F_1(\Sigma^{-1})$	Mean	10	0.78/0.76	0.54/0.23	0.67/0.58	0.3/0.57	0.4/NA
$F_1(B)/F_1(\Sigma^{-1})$	Mean	20	0.76/0.75	0.49/0.21	0.48/0.49	0.27/0.57	0.39/NA
$F_1(B)/F_1(\Sigma^{-1})$	Cov	5	0.77/0.75	0.58/0.3	0.59/0.58	0.48/0.68	0.39/NA
$F_1(B)/F_1(\Sigma^{-1})$	Cov	10	0.77/0.75	0.64/0.29	0.57/0.57	0.42/0.67	0.39/NA
$F_1(B)/F_1(\Sigma^{-1})$	Cov	20	0.78/0.73	0.38/0.23	0.49/0.47	0.38/0.48	0.37/NA

first 26 weeks as training set, and log-returns of the remaining 26 weeks as testing set. The tuning parameters were selected using the modified 10-fold cross-validation described in Section 2.4. Table 3.2 reports the mean squared prediction error (MSPE) of the five comparison methods. Even though all methods are competitive on this dataset, REG-ISE estimator achieves the smallest prediction error. Figure 3.3 presents the graphs induced by the estimates of $\mathbf{\Sigma}^{-1}$ using MRCE and REG-ISE, respectively. Comparing the two graphs, both MRCE and REG-ISE indicate that companies from the same industry are partially correlated, e.g. GE and IBM (technology), Ford and GM (auto industry). AIG (insurance company) seems to be partially correlated with most of the other companies. However, there are some discrepancies between the two graphs, e.g., GM is found to be partially correlated to IBM by REG-ISE but to be uncorrelated by MRCE. Overall, the results from REG-ISE have reasonable financial interpretation.

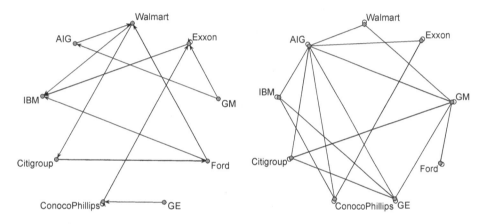

Figure 3.3: Graphs from the estimates of the inverse covariance matrix $\mathbf{\Sigma}^{-1}$ obtained by MRCE (left) and REG-ISE(right).

Table 3.2: Prediction accuracy measured by MSPE for various methods for the asset return dataset.

Method	MSPE
REG-ISE	**0.69 ± 0.11**
MRCE	0.71 ± 0.12
l_1-CGGM	0.72 ± 0.10
CAPME	0.72 ± 0.11
LAD-Lasso	0.73 ± 0.12

3.5.2 eQTL Data Analysis

We analyze yeast eQTL dataset (6) which contains genotype data for 2,956 SNPs (predictors) and microarray data for 6,216 genes (responses) regarding 112 segregants (instances). We extracted 1,260 unique SNPs, and focused on 125 genes belonging to cell-cycle pathway provided by the KEGG database (17). For all methods, the tuning parameters were chosen via 5-fold modified cross-validation described in Section 3.3.5. We first evaluated the predictive accuracy of our method and the comparison methods by randomly partitioning the data into training and test sets, using 90 observations for training and the remainder for testing. We computed the MSPE for the testing set. The average MSPEs based on 20 random partitions are presented in Table 3.3. We can see that overall the predictive performance of REG-ISE is superior to the other methods.

Figure 3.4 shows the cell-cycle pathways estimated by the proposed REG-ISE, MRCE method, CAPME method along with the benchmark KEGG pathway. MRCE tends to estimate many spurious links. Similar observation holds for l_1-CGGM, so its estimated graph is not represented here. CAPME recovers some of the links but not as accurately as REG-ISE. This can be partly explained by the fact that CAPME does not take into account the covariance structure in its regression stage and does not have any feedback loop. This can result in poor estimation of the regression matrix $\boldsymbol{B}$, which in turn may negatively impact the estimation of precision matrix $\boldsymbol{\Sigma}^{-1}$. In addition, lack of robustness can also result in inaccurate network reconstruction. Certain discrepancies between true and estimated graphs may also be caused by inherent limitations in this dataset. For instance, some edges in cell-cycle pathway may not be observable from gene expression data. Additionally in this dataset, the perturbation of cellular systems may not be significant enough to enable accurate inference of some of the links.

Using the KEGG pathway as the "ground truth", we also computed the F_1 scores for the estimates of $\boldsymbol{\Sigma}^{-1}$ shown in Table 3.4. As a sanity check, we analyzed the microarray data without SNPs as predictors using glasso. The resulting graph was extremely dense with F_1 score to be 0.033. This indicates the disadvantage of procedures like glasso which is unable to adjust

Table 3.3: MSPEs under different methods based on 20 random partitions of the eQTL into training and test sets.

REG-ISE	MRCE	l_1-CGGM	CAPME
2.36 ± 0.07	6.25 ± 0.22	4.46 ± 0.17	4.38 ± 0.09

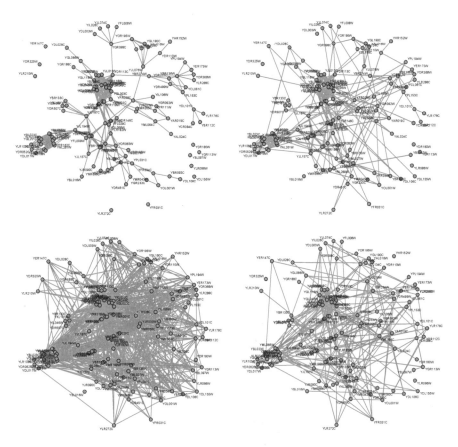

Figure 3.4: Yeast cell cycle network provided in the KEGG database (top left), estimated by REG-ISE (top right), MRCE (bottom left), and CAPME (bottom right).

Table 3.4: F_1 scores of the estimated cell-cycle network (higher values indicate higher accuracy).

REG-ISE	MRCE	l_1-CGGM	CAPME	Glasso
0.635	0.042	0.089	0.348	0.033
±0.009	±0.008	±0.011	±0.034	±0.014

for predictors (hereby the genetic variants) in inverse covariance matrix estimation.

From reconstructed network and F_1 scores, we conclude that REG-ISE most faithfully estimates the cell-cycle network compared to the other methods, which clearly demonstrates the value of embracing the robustness.

3.6   Concluding Remarks

In this chapter, we developed a robust framework to jointly estimate multiresponse regression and inverse covariance matrix from high dimensional data.
Our framework yields log-non-linear formulations can non-convex problems,
which is the price to pay to significantly improve robustness with respect
to noise and outliers. We employed the traditional parametrization for the
multiresponse linear regression model, namely we used $(\boldsymbol{B}, \boldsymbol{\Sigma})$ directly. This
parametrization allowed us to gain clarity and insights on the robusteness
of our formulation. It also enables the straightforward applicability of our
framework to single regression and to inverse covariance estimation on its
own. As future work we plan to apply our framework to the parametrization
proposed by (28), which is based on the covariance of the joint distribution for predictors and responses. The proposed methodology is valuable for
many applications beyond the integration of genomic and transcriptomic
data and financial data analysis. Additional interesting future work includes extending the proposed method to directed graph modeling via vector
autoregressive models, and extending our theoretical results to the high-
dimensional setting.

3.7   Appendix: Proof of Theorem 1

Recall our objective function:

$$L_{n,\lambda}(\boldsymbol{B}, \boldsymbol{\Sigma}^{-1}) =$$
$$- \log \left[\frac{1}{n} \sum_{i=1}^{n} \exp(-\frac{1}{2}(\boldsymbol{y}_i - \boldsymbol{B}'\boldsymbol{x}_i)'\boldsymbol{\Sigma}^{-1}(\boldsymbol{y}_i - \boldsymbol{B}'\boldsymbol{x}_i)) \right]$$
$$- \frac{1}{2} \log |\boldsymbol{\Sigma}^{-1}| + \lambda_1 \|\boldsymbol{\Sigma}^{-1}\|_1 + \lambda_2 \|\boldsymbol{B}\|_1 \qquad (3.12)$$

Let $\bar{\boldsymbol{B}}$ and $\bar{\boldsymbol{\Sigma}}^{-1}$ be the true regression and inverse covariance matrices. We follow the same reasoning as in the proof of Theorem 1 in (12). The key idea is that it's enough to show that for any $\delta > 0$ there exists a large constant C, such that

$$P\{ \sup_{\|\boldsymbol{U}\|=C} L_{n,\lambda}(\bar{\boldsymbol{B}} + \frac{\boldsymbol{U}_1}{\sqrt{n}}, \bar{\boldsymbol{\Sigma}}^{-1} + \frac{\boldsymbol{U}_2}{\sqrt{n}}) > L_{n,\lambda}(\bar{\boldsymbol{B}}, \bar{\boldsymbol{\Sigma}}^{-1})\} \geq 1 - \delta$$

with $\boldsymbol{U} = (vec(\boldsymbol{U}_1)', vec(\boldsymbol{U}_2)')'$.
Define $Q(\boldsymbol{B}, \boldsymbol{\Sigma}^{-1})$ as

$$Q(\boldsymbol{B}, \boldsymbol{\Sigma}^{-1}) = - \log \left[\frac{1}{n} \sum_{i=1}^{n} \exp(-\frac{1}{2}(\boldsymbol{y}_i - \boldsymbol{B}'\boldsymbol{x}_i)'\boldsymbol{\Sigma}^{-1}(\boldsymbol{y}_i - \boldsymbol{B}'\boldsymbol{x}_i)) \right].$$

By using a similar reasoning as in Section 3.3.4, one can show that around the true parameters $(\bar{\boldsymbol{B}}, \bar{\boldsymbol{\Sigma}}^{-1})$, the difference $Q(\bar{\boldsymbol{B}} + \frac{\boldsymbol{U}_1}{\sqrt{n}}, \bar{\boldsymbol{\Sigma}}^{-1} + \frac{\boldsymbol{U}_2}{\sqrt{n}}) - Q(\bar{\boldsymbol{B}}, \bar{\boldsymbol{\Sigma}}^{-1})$ can be lower-bounded by

$$\frac{1}{n} \sum_{i=1}^{n} w_i (\boldsymbol{y}_i - (\bar{\boldsymbol{B}} + \frac{\boldsymbol{U}_1}{\sqrt{n}})'\boldsymbol{x}_i)'(\bar{\boldsymbol{\Sigma}}^{-1} + \frac{\boldsymbol{U}_2}{\sqrt{n}})(\boldsymbol{y}_i - (\bar{\boldsymbol{B}} + \frac{\boldsymbol{U}_1}{\sqrt{n}})'\boldsymbol{x}_i)$$
$$- \frac{1}{n} \sum_{i=1}^{n} w_i (\boldsymbol{y}_i - \bar{\boldsymbol{B}}'\boldsymbol{x}_i)'\boldsymbol{\Sigma}^{-1}(\boldsymbol{y}_i - \bar{\boldsymbol{B}}'\boldsymbol{x}_i) + o(1)$$

where $w_i \equiv w_i(\boldsymbol{\beta}_0) = \frac{\exp(g_i(\boldsymbol{\beta}_0))}{\frac{1}{n}\sum_{i=1}^{n}\exp(g_i(\boldsymbol{\beta}_0))}$. Even though Q is globally non-convex, such an approximation is valid as Q is locally bi-convex in a neighborhood of the true parameters $(\bar{\boldsymbol{B}}, \bar{\boldsymbol{\Sigma}}^{-1})$ with asymptotic probability one. Briefly, one can show that the events $\frac{\boldsymbol{U}_1^T}{\sqrt{n}}\nabla_B^2 Q(\bar{\boldsymbol{B}} + t\frac{\boldsymbol{U}_1}{\sqrt{n}})\frac{\boldsymbol{U}_1}{\sqrt{n}} < 0$ and $\frac{\boldsymbol{U}_2^T}{\sqrt{n}}\nabla_{\Sigma^{-1}}^2 Q(\bar{\boldsymbol{\Sigma}}^{-1} + t\frac{\boldsymbol{U}_2}{\sqrt{n}})\frac{\boldsymbol{U}_2}{\sqrt{n}} < 0$ for $t \in (0,1)$ are small uniformly on $\boldsymbol{U}_1$ and $\boldsymbol{U}_2$ as long as $\frac{\|\boldsymbol{U}\|}{\sqrt{n}}$ is small enough. Briefly, this can be done by 'brute force' calculation of the Hessians around the true solution, noticing that the expected value of $\frac{\boldsymbol{U}_1^T}{\sqrt{n}}\nabla_B^2 Q(\bar{\boldsymbol{B}} + t\frac{\boldsymbol{U}_1}{\sqrt{n}})\frac{\boldsymbol{U}_1}{\sqrt{n}}$ and $\frac{\boldsymbol{U}_2^T}{\sqrt{n}}\nabla_{\Sigma^{-1}}^2 Q(\bar{\boldsymbol{\Sigma}}^{-1} + t\frac{\boldsymbol{U}_2}{\sqrt{n}})\frac{\boldsymbol{U}_2}{\sqrt{n}}$ are both non-negative for $\frac{\|U\|}{\sqrt{n}}$ small enough and then upper-bounding the large deviations from the expected values using Azuma-Hoeffding's inequality.

Let us define $V_n(U)$ as

$$V_n(U) = L_{n,\lambda}(\bar{B} + \frac{U_1}{\sqrt{n}}, \bar{\Sigma}^{-1} + \frac{U_2}{\sqrt{n}}) - L_{n,\lambda}(\bar{B}, \bar{\Sigma}^{-1}).$$

For notation convenience, denote $\tilde{Y} = \mathrm{diag}(\sqrt{w_1}, \ldots, \sqrt{w_n})Y$, and $\tilde{X} = \mathrm{diag}(\sqrt{w_1}, \ldots, \sqrt{w_n})X$. Noting that $|\bar{B}_{jk} + \frac{u_{1j,k}}{\sqrt{n}}| - |\bar{B}_{jk}| = |\frac{u_{1j,k}}{\sqrt{n}}|$ for $\bar{B}_{jk} = 0$ and $|\bar{\Sigma}_{st}^{-1} + \frac{u_{2s,t}}{\sqrt{n}}| - |\bar{\Sigma}_{st}^{-1}| = |\frac{u_{2s,t}}{\sqrt{n}}|$ for $\bar{\Sigma}_{st}^{-1} = 0$, we then have

$$\begin{aligned}
V_n(U) \geq &-\log|(\bar{\Sigma}^{-1} + \frac{U_2}{\sqrt{n}})\bar{\Sigma}| \\
&+ \frac{1}{n}\mathrm{trace}\{(\bar{\Sigma}^{-1} + \frac{U_2}{\sqrt{n}})(\tilde{Y} - \tilde{X}(\bar{B} + \frac{U_1}{\sqrt{n}}))'(\tilde{Y} - \tilde{X}(\bar{B} + \frac{U_1}{\sqrt{n}}))\} \\
&- \frac{1}{n}\mathrm{trace}\{\bar{\Sigma}^{-1}(\tilde{Y} - \tilde{X}\bar{B})'(\tilde{Y} - \tilde{X}\bar{B})\} \\
&+ \lambda_1 \sum_{B_{kj \neq 0}} (|\bar{B}_{jk} + \frac{u_{1j,k}}{\sqrt{n}}| - |\bar{B}_{jk}|) \\
&+ \lambda_2 \sum_{\bar{\Sigma}_{st \neq 0}^{-1}} (|\bar{\Sigma}_{st}^{-1} + \frac{u_{2s,t}}{\sqrt{n}}| - |\bar{\Sigma}_{jk}^{-1}|).
\end{aligned}$$

Following the same derivations as the proof of Lemma 3 in (19), we can show that for a sufficiently large C, $V_n(U) > 0$ uniformly on $\{U : \|U\| = C\}$ with probability greater than $1 - \delta$. This completes the proof.

References

1. O. Banerjee, L. Ghaoui, and A. d'Aspremont. Model selection through sparse maximum likelihood estimation. *JMLR*, 9:485–516, 2008.

2. A. Basu, I. R. Harris, N. L. Hjort, and M. C. Jones. Robust and efficient estimation by minimising a density power divergence. *Biometrika*, 85, 1998.

3. R. Beran. Robust location estimates. *Annals of Statistics*, 5:431–444, 1977.

4. P. J. Bickel and E. Levina. Regularized estimation of large covariance matrices. *Annals of Statistics*, 36(1):199–227, 2008.

5. L. Breiman and J. H. Friedman. Predicting multivariate responses in multiple linear regression. *JRSS Series B.*, 1997.

6. R. Brem and L. Kruglyak. The landscape of genetic complexity across

5,700 gene expression traits in yeast. *Proc. Natl. Acad. Sci. USA*, 102(5):1572–1577, 2005.

7. T. T. Cai, H. Li, W. Liu, and J. Xie. Covariate adjusted precision matrix estimation with an application in genetical genomics. *Biometrika*, 2011.

8. E. Candes and T. Tao. The dantzig selector: Statistical estimation when p is much larger than n. *The Annals of Statistics*, 35:2313–2351, 2007.

9. F. E. Curtis and M. L. Overton. A Sequential Quadratic Programming Algorithm for Nonconvex, Nonsmooth Constrained Optimization. *SIAM Journal on Optimization*, 22(2):474–500, 2011.

10. F. E. Curtis and X. Que. An Adaptive Gradient Sampling Algorithm for Nonsmooth Optimization. *Optimization Methods and Software*, 2011.

11. D. L. Donoho and R. C. Liu. The "automatic" robustness of minimum distance functional. *Annals of Statistics*, 16:552–586, 1994.

12. J. Fan and R. Li. Variable selection via nonconcave penalized likelihood and its oracle properties. *JASA*, 96:1348–1360, 2001.

13. M. Finegold and M. Drton. Robust graphical modeling of gene networks using classical and alternative t-distribution. *Annals of Applied Statistics*, 5(2A):1075–1080, 2011.

14. J. Friedman, T. Hastie, and R. Tibshirani. Sparse inverse covariance estimation with the graphical lasso. *Biostatistics*, 9(3):432–441, 2008.

15. C. Hsieh, M. Sustik, I. Dhillon, and P. Ravikumar. Sparse inverse covariance matrix estimation using quadratic approximation. In *NIPS*, 2011.

16. J. Huang, N. Liu, M. Pourahmadi, and L. Liu. Covariance matrix selection and estimation via penalised normal likelihood. *Biometrika*, 93(1):85–98, 2006.

17. M. Kanehisa, S. Goto, M. Furumichi, M. Tanabe, and M. Hirakawa. Kegg for representation and analysis of molecular networks involving diseases and drugs. *Nucleic Acid Research*, 31(1):355–360, 2010.

18. S. L. Lauritzen. *Graphical Models*. Oxford: Clarendon Press, 1996.

19. W. Lee and Y. Liu. Simultaneous multiple response regression and inverse covariance matrix estimation via penalized gaussian maximum likelihood. *J. Multivar. Anal.*, 2012.

20. E. Levina, A. J. Rothman, and J. Zhu. Sparse estimation of large covariance matrices via a nested lasso penalty. *Annals of Applied Statistics*, 2(1), 2008.

21. N. Meinshausen and P. Buhlmann. High-dimensional graphs and variable selection with the lasso. *Annals of Statistics*, 34(3):1436–1462, 2006.

22. S. Negahban, P. Ravikumar, M. Wainwright, and B. Yu. A unified framework for high-dimensional analysis of M-estimators with decomposable regularizers. *CoRR*, abs/1010.2731, 2010.

23. G. Obozinski, B. Taskar, and M. Jordan. Multi-task feature selection. Tech. report, University of California, Berkeley, 2006.

24. P. Z. G. Qian and C. F. J. Wu. Sliced space-filling designs. *Biometrika*, 96(4):945–956, 2009.

25. P. Ravikumar, M. Wainwright, G. Raskutti, and B. Yu. High-dimensional covariance estimation by minimizing l1-penalized log-determinant divergence. *Electron. J. Statist.*, 5:935–980, 2011.

26. A. Rothman, E. Levina, and J. Zhu. Sparse multivariate regression with covariance estimation. *JCGS*, 19(4):947–962, 2010.

27. D. Scott. Parametric statistical modeling by minimum integrated square error. *Technometrics*, 43:274–285, 2001.

28. K. Sohn and S. Kim. Joint estimation of structured sparsity and output structure in multiple-output regression via inverse-covariance regularization. *AISTATS*, 2012.

29. M. Sugiyama, T. Suzuki, T. Kanamori, M. C. Du Plessis, S. Liu, and I. Takeuchi. Density-difference estimation. *NIPS*, 25:692–700, 2012.

30. R. Tibshirani. Regression shrinkage and selection via the lasso. *Journal of the Royal Statistical Society, Series B*, 58(1):267–288, 1996.

31. H. Wang, G. Li, and G. Jiang. Robust regression shrinkage and consistent variable selection through the LAD-lasso. *J. Business and Economics Statistics*, 25:347–355, 2007.

32. J. Wolfowitz. The minimum distance method. *Annals of Mathematical Statistics*, 28(1):75–88, 1957.

33. M. Yuan and Y. Lin. Model selection and estimation in the gaussian graphical model. *Biometrika*, 94(1):19–35, 2007.

4   Semistochastic Quadratic Bound Methods

Aleksandr Aravkin
IBM T. J. Watson Research Center
Yorktown Heights, NY, USA

saravkin@us.ibm.com

Anna Choromanska
Courant Institute, NYU
New York, NY, USA

achoroma@cims.nyu.edu

Tony Jebara
Columbia University
New York, NY, USA

jebara@cs.columbia.edu

Dimitri Kanevsky
Google INC
New York, NY, USA

dkanevsky@google.com

In this chapter we focus on partition function-based optimization. Partition functions arise in a variety of settings, including conditional random fields, logistic regression, and latent gaussian models. In this chapter, we consider semistochastic quadratic bound (SQB) methods for maximum likelihood inference based on partition function optimization. Batch methods based on the quadratic bound were recently proposed for this class of problems, and performed favorably in comparison to state-of-the-art techniques. Semistochastic methods fall in between batch algorithms, which use all the data, and stochastic gradient type methods, which use small random selections at each iteration. We build semistochastic quadratic bound-based methods, and prove both global convergence (to a stationary point) under very weak assumptions, and linear convergence rate under stronger assumptions on the objective. To make the proposed methods faster and more stable, we consider inexact subproblem minimization and batch-size selection schemes. The efficacy of SQB methods is demonstrated via comparison with several state-of-the-art techniques on commonly used datasets.

4.1   Introduction

The problem of optimizing a cost function expressed as the sum of a loss term over each sample in an input dataset is pervasive in machine learning. One example of a cost function of this type is the partition funtion, which is studied in this chapter. It is a central quantity in many different learning tasks including training conditional random fields (CRFs) and log-linear models (11). Batch methods based on quadratic bounds were recently proposed (11) for a class of problems invoving the minimization of the partition function, and performed favorably in comparison to state-of-the-art techniques. This chapter focuses on a semistochastic extension of this recently developed optimization method. Standard learning systems based on batch methods such as BFGS and memory-limited L-BFGS, steepest descent (see e.g. (22)), conjugate gradient (10), or quadratic bound majorization methods (11) need to make a full pass through an entire dataset before updating the parameter vector. Even though these methods can converge quickly (sometimes in several passes through the dataset), as datasets grow in size, this learning strategy becomes increasingly inefficient. To faciliate learning on massive datasets, the community increasingly turns to stochastic methods.

Stochastic optimization methods interleave the update of parameters after only processing a small mini-batch of examples (potentially as small as a single data-point), leading to significant computational savings (5, 16, 26). Due to its simplicity and low computational cost, the most popular contemporary stochastic learning technique is stochastic gradient descent (SGD) (25, 5, 6). SGD updates the parameter vector using the gradient of the objective function as evaluated on a single example (or, alternatively, a small mini-batch of examples). This algorithm admits multiple extensions, including (i) the stochastic average gradient method (SAG), which averages the most recently computed gradients for each training example (27); (ii) methods that compute the (weighted) average of all previous gradients (23, 35); (iii) averaged stochastic gradient descent methods (ASGD), which compute a running average of parameters obtained by SGD (24); (iv) stochastic dual coordinate ascent, which optimizes the dual objective with respect to a single dual vector or a mini-batch of dual vectors chosen uniformly at random (31, 33); (v) variance reduction techniques (12, 37, 27, 32) (some do not require storage of gradients, c.f. (12)); (vi) majorization-minimization techniques that minimize a majoring surrogate of an objective function (17, 18); and (vii) gain adaptation techniques (28, 29).

Semistochastic methods can be viewed as an interpolation between the expensive reliable updates used by full batch methods, and inexpensive noisy updates used by stochastic methods. They inherit the best of both worlds by approaching the solution more quickly when close to the optimum (like a full batch method) while simultaneously reducing the computational complexity per iteration (though less aggressively than stochastic methods). Several semistochastic extensions have been explored in previous works (38, 34, 13). Recently, convergence theory and sampling strategies for these methods have been explored (9, 8) and linked to results in finite sampling theory (2).

Additionally, incorporating second-order information, by means of namely the Hessian, into the optimization problem (28, 30, 15, 4, 19, 7) has been shown to often improve the performance of traditional SGD methods. These methods typically provide fast improvement initially, but are slow to converge near the optimum (see e.g. (9)), require step-size tuning and are difficult to parallelize (14). This chapter focuses on semistochastic extension of a recently developed quadratic bound majorization technique (11), and we call the new algorithm *semistochastic quadratic bound* (SQB) method. The bound computes the update on the parameter vector using the product of the gradient of the objective function and an inverse of a second-order term that is a descriptor of the curvature of the objective function (different than the Hessian). We discuss implementation details, in particular curvature approximation, inexact solvers, and batch-size selection strategies, which make the running time of our algorithm comparable to the gradient methods and also make the method easily parallelizable. We show global convergence of the method to a stationary point under very weak assumptions (in particular convexity is not required) and then, following the techniques of (9), a linear convergence rate when the size of the mini-batch grows sufficiently fast. This rate of convergence matches state-of-the-art incremental techniques (12, 31, 27, 17) and is better than in case of standard stochastic gradient methods (25, 6), which typically have sublinear convergence rate (27, 1). Compared to other existing majorization-minimization incremental techniques (17), our approach uses much tighter bounds, which, as shown in (11), can lead to faster convergence.

The chapter is organized as follows: Section 4.2 reviews the quadratic bound majorization technique. Section 4.3 discusses stochastic and semistochastic extensions of the bound, and presents convergence theory for the proposed methods. In particular, we discuss very general stationary convergence theory under very weak assumptions, and also present a much stronger theory, including convergence rate analysis, for logistic regression. Section 4.4 discusses implementation details, and Section 4.5 shows numerical experiments illustrating the use of the proposed methods for l_2-regularized logistic regression problems. Section 4.6 contains conclusions.

The semistochastic quadratic bound majorization technique that we develop in this chapter can be broadly applied to mixture models or models that induce representations. The advantages of this technique in the batch setting for learning mixture models and other latent models, was shown in the work of (11). In particular, quadratic bound majorization was able to find better local optima in non-convex problems than state-of-the art methods (and in less time). While theoretical guarantees for non-convex problems are hard to obtain, the broader convergence theory developed in this chapter (finding a stationary point under weak assumptions) does carry over to the non-convex setting.

4.2   Quadratic Bound Methods

Let Ω be a discrete probability space over a set of n elements, and take any log-linear density model

$$p(y|x_j, \boldsymbol{\theta}) = \frac{1}{Z_{x_j}(\boldsymbol{\theta})} h_{x_j}(y) \exp\left(\boldsymbol{\theta}^\top \mathbf{f}_{x_j}(y)\right) \tag{4.1}$$

parametrized by a vector $\boldsymbol{\theta} \in \mathbb{R}^d$, where $\{(x_1, y_1), \dots, (x_T, y_T)\}$ are iid input-output pairs, $\mathbf{f}_{x_j} : \Omega \mapsto \mathbb{R}^d$ is a continuous vector-valued function mapping, and $h_{x_j} : \Omega \mapsto \mathbb{R}^+$ is a fixed non-negative measure. The *partition function* $Z_{x_j}(\boldsymbol{\theta})$ is a scalar that ensures that $p(y|x_j, \boldsymbol{\theta})$ is a true density, so in particular (4.1) integrates to 1:

$$Z_{x_j}(\boldsymbol{\theta}) = \sum_y h_{x_j}(y) \exp(\boldsymbol{\theta}^\top \mathbf{f}_{x_j}(y)). \tag{4.2}$$

There exists a fast method to find a tight quadratic bound for $Z_{x_j}(\boldsymbol{\theta})$ (11). It is shown in the subroutine Bound Computation in Algorithm 4.1, which finds $z, \mathbf{r}, \mathbf{S}$ so that

$$Z_{x_j}(\boldsymbol{\theta}) \leq z \exp(\tfrac{1}{2}(\boldsymbol{\theta} - \tilde{\boldsymbol{\theta}})^\top \mathbf{S}(\boldsymbol{\theta} - \tilde{\boldsymbol{\theta}}) + (\boldsymbol{\theta} - \tilde{\boldsymbol{\theta}})^\top \mathbf{r}) \tag{4.3}$$

for any $\boldsymbol{\theta}, \tilde{\boldsymbol{\theta}}, \mathbf{f}_{x_j}(y) \in \mathbb{R}^d$ and $h_{x_j}(y) \in \mathbb{R}^+$ for all $y \in \Omega$.

The (regularized) maximum likelihood estimation problem is equivalent to

$$\min_{\boldsymbol{\theta}} \left\{ \mathcal{L}_\eta(\boldsymbol{\theta}) := -\frac{1}{T} \sum_{j=1}^T \log(p(y_j|x_j, \boldsymbol{\theta})) + \frac{\eta}{2}\|\boldsymbol{\theta}\|^2 \right.$$

$$\left. \approx \frac{1}{T} \sum_{j=1}^T \left(\log(Z_{x_j}(\boldsymbol{\theta})) - \boldsymbol{\theta}^\top \mathbf{f}_{x_j}(y_j) \right) + \frac{\eta}{2}\|\boldsymbol{\theta}\|^2 \right\}, \tag{4.4}$$

where $\approx$ means equal up to an additive constant. The bound (4.3) suggests the iterative minimization scheme (derived in (11)):

$$\boldsymbol{\theta}^{k+1} = \boldsymbol{\theta}^k - \alpha_k(\boldsymbol{\Sigma}^k + \eta\boldsymbol{I})^{-1}(\boldsymbol{\mu}^k + \eta\boldsymbol{\theta}^k). \tag{4.5}$$

where $\boldsymbol{\Sigma}^k$ and $\boldsymbol{\mu}^k$ are computed using Algorithm 4.1, η is the regularization term and α_k is the step size at iteration k.

In this chapter, we consider applying the bound to randomly selected batches of data; any such selection we denote $\mathcal{T} \subset [1, \ldots, T]$ or $\mathcal{S} \subset [1, \ldots, T]$.

Algorithm 4.1 Semistochastic Quadratic Bound (SQB)

Input Parameters $\tilde{\boldsymbol{\theta}}, \mathbf{f}_{x_j}(y) \in \mathbb{R}^d$ and $h_{x_j}(y) \in \mathbb{R}^+$ for $y \in \Omega$, $j \in \mathcal{T}$

Initialize $\boldsymbol{\mu}_{\mathcal{T}} = \mathbf{0}, \boldsymbol{\Sigma}_{\mathcal{T}} = \mathbf{0}(d, d)$

For each $j \in \mathcal{T}$

    Subroutine **Bound Computation:**

    $z \to 0^+, \mathbf{r} = \mathbf{0}, \mathbf{S} = z\mathbf{I}$

    For each $y \in \Omega$

        $\alpha = h_{x_j}(y)\exp(\tilde{\boldsymbol{\theta}}^\top \mathbf{f}_{x_j}(y))$

        $\mathbf{S} += \frac{\tanh(\frac{1}{2}\log(\alpha/z))}{2\log(\alpha/z)}(\mathbf{f}_{x_j}(y) - \mathbf{r})(\mathbf{f}_{x_j}(y) - \mathbf{r})^\top$

        $\mathbf{r} = \frac{z}{z+\alpha}\mathbf{r} + \frac{\alpha}{z+\alpha}\mathbf{f}_{x_j}(y)$

        $z += \alpha$

    Subroutine output $z, \mathbf{r}, \mathbf{S}$

    $\boldsymbol{\mu}_{\mathcal{T}} += \mathbf{r} - \mathbf{f}_{x_j}(y)$

    $\boldsymbol{\Sigma}_{\mathcal{T}} += \mathbf{S}$

$\boldsymbol{\mu}_{\mathcal{T}} /= |\mathcal{T}|$

$\boldsymbol{\Sigma}_{\mathcal{T}} /= |\mathcal{T}|$

Output $\boldsymbol{\mu}_{\mathcal{T}}, \boldsymbol{\Sigma}_{\mathcal{T}}$

4.3 Stochastic and Semistochastic Extensions

The bounding method proposed in (11) is summarized in Algorithm 4.1 with $\mathcal{T} = [1, \ldots, T]$ at every iteration. When T is large, this strategy can be expensive. In fact, computing the bound has complexity $O(Tnd^2)$, since Tn outer products must be summed to obtain $\boldsymbol{\Sigma}$, and each other product has complexity $O(d^2)$. When the dimension d is large, considerable speedups can be gained by obtaining a factored form of $\boldsymbol{\Sigma}$, as described in Section 4.4.1. Nonetheless, in either strategy, the size of T is a serious issue.

A natural approach is to subsample a smaller selection $\mathcal{T}$ from the training set $[1, \dots, T]$, so that at each iteration, we run Algorithm 4.1 over $\mathcal{T}$ rather than over the full data to get $\boldsymbol{\mu}_{\mathcal{T}}, \boldsymbol{\Sigma}_{\mathcal{T}}$. When $|\mathcal{T}|$ is fixed (and smaller than T), we refer to the resulting method as a *stochastic* extension. If instead $|\mathcal{T}|$ is allowed to grow as iterations proceed, we call this method *semistochastic*; these methods are analyzed in (9). All of the numerical experiments we present focus on semistochastic methods. One can also decouple the computation of gradient and curvature approximations, using different data selections (which we call $\mathcal{T}$ and $\mathcal{S}$). We show that this development is theoretically justifiable and practically very useful.

For the stochastic and semistochastic methods discussed here, the quadratic bound property (4.3) does not hold for $Z(\boldsymbol{\theta})$, so the convergence analysis of (11) does not immediately apply. Nonetheless, it is possible to analyze the algorithm in terms of sampling strategies for $\mathcal{T}$.

The appeal of the stochastic modification is that when $|\mathcal{T}| \ll T$, the complexity $O(|\mathcal{T}|nd)$ of Algorithm 4.1 to compute $\boldsymbol{\mu}_{\mathcal{T}}, \boldsymbol{\Sigma}_{\mathcal{S}}$ is much lower; and then we can still implement a (modified) iteration (4.5). Intuitively, one expects that even small samples from the data can give good updates for the overall problem. This intuition is supported by experimental results, which show that in terms of effective passes through the data, SQB is competitive with state-of-the-art methods.

We now present the theoretical analysis of Algorithm 4.1. We first prove that under very weak assumption, in particular using only the Lipschitz property, but not requiring convexity of the problem, the proposed algorithm converges to a stationary point. The proof technique easily carries over to other objectives, such as the ones used in maximum latent conditional likelihood problems (for details see (11)), since it relies mainly on the sampling method used to obtain $\mathcal{T}$. Then we focus on problem (4.4), which is convex, and strictly convex under appropriate assumptions on the data. We use the structure of (4.4) to prove much stronger results, and in particular analyze the rate of convergence of Algorithm 4.1.

4.3.1 General Convergence Theory

We present a general global convergence theory, that relies on the Lipschitz property of the objective and on the sampling strategy in the context of Algorithm 4.1. The end result we show here is that any limit point of the iterates is *stationary*. We begin with two simple preliminary results.

Lemma 4.1. *If every $i \in [1, \dots, T]$ is equally likely to appear in $\mathcal{T}$, then* $E[\boldsymbol{\mu}_{\mathcal{T}}] = \boldsymbol{\mu}$.

Proof. Algorithm 4.1 returns $\boldsymbol{\mu}_{\mathcal{T}} = \frac{1}{|\mathcal{T}|} \sum_{j \in \mathcal{T}} \psi_j(\boldsymbol{\theta})$,
where $\psi_j(\boldsymbol{\theta}) = -\nabla_{\boldsymbol{\theta}} \log(p(y_j|x_j, \boldsymbol{\theta}))$. If each j has an equal chance to appear in $\mathcal{T}$, then

$$E\left[\frac{1}{|\mathcal{T}|} \sum_{j \in \mathcal{T}} \psi_j(\boldsymbol{\theta})\right] = \frac{1}{|\mathcal{T}|} \sum_{j \in \mathcal{T}} E[\psi_j(\boldsymbol{\theta})] = \frac{1}{|\mathcal{T}|} \sum_{j \in \mathcal{T}} \boldsymbol{\mu} = \boldsymbol{\mu}.$$

$\square$

Note that the hypothesis here is very weak: there is no stipulation that the batch size be of a certain size, grow with iterations, etc. This lemma therefore applies to a wide class of randomized bound methods.

Lemma 4.2. *Denote by $\lambda_{\min}$ the infimum over all possible eigenvalues of $\boldsymbol{\Sigma}_{\mathcal{S}}$ over all choices of batches ($\lambda_{\min}$ may be 0). Then $E[(\boldsymbol{\Sigma}_{\mathcal{S}}+\eta\boldsymbol{I})^{-1}]$ satisfies*

$$\frac{1}{\eta + \lambda_{\max}} \boldsymbol{I} \le E[(\boldsymbol{\Sigma}_{\mathcal{S}} + \eta \boldsymbol{I})^{-1}] \le \frac{1}{\eta + \lambda_{\min}} \boldsymbol{I}.$$

Proof. For any vector $\mathbf{x}$ and any realization of $\boldsymbol{\Sigma}_{\mathcal{S}}$, we have

$$\frac{1}{\eta + \lambda_{\max}} \|\mathbf{x}\|^2 \le \mathbf{x}^T (\boldsymbol{\Sigma}_{\mathcal{S}} + \eta \boldsymbol{I})^{-1} \mathbf{x} \le \frac{1}{\eta + \lambda_{\min}} \|\mathbf{x}\|^2 ,$$

where $\lambda_{\max}$ depends on the data. Taking the expectation over $\mathcal{T}$ of the above inequality gives the result. $\square$

Theorem 4.3. *Consider the iterative scheme of Equation (4.5) for solving the (regularized) maximum likelihood estimation problem formulated in Equation (4.4), where at each iteration $\boldsymbol{\mu}_{\mathcal{T}}, \boldsymbol{\Sigma}_{\mathcal{S}}$ is obtained by Algorithm 4.1 for two independently drawn batches subsets $\mathcal{T}, \mathcal{S} \subset [1, \ldots, T]$ selected to satisfy the assumptions of Lemma 4.1. Assume that the step sizes α_k are square summable but not summable. Then $\mathcal{L}_\eta(\boldsymbol{\theta}^k)$ converges to a finite value, and $\nabla \mathcal{L}_\eta(\boldsymbol{\theta}^k) \to 0$. Furthermore, every limit point of $\boldsymbol{\theta}^k$ is a stationary point of $\mathcal{L}_\eta$.*

Theorem 4.3 states the conclusions of (3, Proposition 3), and so to prove it we need only check that the hypotheses of this proposition are satisfied.

Proof. (3) consider algorithms of the form

$$\boldsymbol{\theta}^{k+1} = \boldsymbol{\theta}^k - \alpha_k(\boldsymbol{s}^k + \boldsymbol{w}^k) .$$

In the context of iteration (4.5), at each iteration we have

$$\boldsymbol{s}^k + \boldsymbol{w}^k = (\boldsymbol{\Sigma}_{\mathcal{S}}^k + \lambda \boldsymbol{I})^{-1} \boldsymbol{g}_{\mathcal{T}}^k,$$

where $g_{\mathcal{T}}^k = \mu_{\mathcal{T}}^k + \eta\theta^k$, and g^k is the full gradient of the regularized problem (4.4). We choose

$$s^k = E[(\Sigma_{\mathcal{S}}^k + \eta I)^{-1}]g^k, \quad w^k = (\Sigma_{\mathcal{S}}^k + \eta I)^{-1}g_{\mathcal{T}}^k - s^k.$$

We now have the following results:

1. Unbiased error:

$$\begin{aligned}
E[w^k] &= E[(\Sigma_{\mathcal{S}}^k + \eta I)^{-1}g_{\mathcal{T}}^k - s^k] \\
&= E[(\Sigma_{\mathcal{S}}^k + \eta I)^{-1}]E[g_{\mathcal{T}}^k] - s^k = 0 \ ,
\end{aligned} \tag{4.6}$$

where the second equality is obtained by independence of the batches $\mathcal{T}$ and $\mathcal{S}$, and the last equality uses Lemma 4.1.

2. Gradient related condition:

$$(g^k)^T s^k = (g^k)^T E[(\Sigma_{\mathcal{S}}^k + \eta I)^{-1}]g^k \ \geq \ \frac{\|g^k\|^2}{\eta + \lambda_{\max}}. \tag{4.7}$$

3. Bounded direction:

$$\|s^k\| \leq \frac{\|g^k\|}{\eta + \lambda_{\min}}. \tag{4.8}$$

4. Bounded second moment:
By part 1, we have

$$\begin{aligned}
E[\|w^k\|^2] &\leq E[\|(\Sigma_{\mathcal{S}}^k + \eta I)^{-1}g_{\mathcal{T}}^k\|^2] \\
&\leq \frac{E[\|g_{\mathcal{T}}^k\|^2]}{(\eta + \lambda_{\min})^2} = \frac{\mathrm{tr} \ (\ \mathrm{cov} \ [g_{\mathcal{T}}^k]) + \|g^k\|^2}{(\eta + \lambda_{\min})^2}.
\end{aligned} \tag{4.9}$$

The covariance matrix of $g_{\mathcal{T}}^k$ is proportional to the covariance matrix of the set of individual (data-point based) gradient contributions, and for problems of form (4.4) these contributions lie in the convex hull of the data, so in particular the trace of the covariance must be finite. Taken together, these results show all hypotheses of (3, Proposition 3) are satisfied, and the result follows.   □

Theorem (4.3) applies to any stochastic and semistochastic variant of the method. Note that two independent data samples $\mathcal{T}$ and $\mathcal{S}$ are required to prove (4.6). Computational complexity motivates different strategies for selecting $\mathcal{T}$ and $\mathcal{S}$. In particular, a strategy to use larger mini-batches to estimate the gradient, and smaller mini-batch sizes for the estimation of the second-order curvature term (for computational efficiency) has been explored in the context of stochastic Hessian methods (7). We describe our implementation details in Section 4.4.

4.3.2 Rates of Convergence for Logistic Regression

The structure of objective (4.4) allows for a much stronger convergence theory. We first present a lemma characterizing strong convexity and the Lipschitz constant for (4.4). Both of these properties are crucial to the convergence theory.

Lemma 4.4. *The objective $\mathcal{L}_\eta$ in (4.4) has a gradient that is uniformly norm bounded, and Lipschitz continuous.*

Proof. The function $\mathcal{L}_\eta$ has a Lipschitz continuous gradient if there exists an L such that

$$\|\nabla \mathcal{L}_\eta(\boldsymbol{\theta}^1) - \nabla \mathcal{L}_\eta(\boldsymbol{\theta}^0)\| \le L \|\boldsymbol{\theta}^1 - \boldsymbol{\theta}^0\|$$

holds for all $(\boldsymbol{\theta}^1, \boldsymbol{\theta}^0)$. Any uniform bound for   trace $(\nabla^2 \mathcal{L}_\eta)$ is a Lipschitz bound for $\nabla \mathcal{L}_\eta$. Define

$$a_{y,j} := h_{x_j}(y) \exp(\boldsymbol{\theta}^\top \mathbf{f}_{x_j}(y)) \, ,$$

and note $a_{y,j} \ge 0$. Let $\mathbf{p}_j$ be the empirical density where the probability of observing y is given by $\dfrac{a_{y,j}}{\sum_y a_{y,j}}$. The gradient of (4.4) is given by

$$\frac{1}{T} \sum_{j=1}^{T} \left(\left(\sum_y \frac{a_{y,j} \mathbf{f}_{x_j}(y)}{\sum_y a_{y,j}} \right) - \mathbf{f}_{x_j}(y_j) \right) + \eta \boldsymbol{\theta}$$

$$= \frac{1}{T} \sum_{j=1}^{T} \left(E_{\mathbf{p}_j}[\mathbf{f}_{x_j}(\cdot)] - \mathbf{f}_{x_j}(y_j) \right) + \eta \boldsymbol{\theta} \tag{4.10}$$

It is straightforward to check that the Hessian is given by

$$\nabla^2 \mathcal{L}_\eta = \frac{1}{T} \sum_{j=1}^{T} cov_{\mathbf{p}_j}[\mathbf{f}_{x_j}(\cdot)] + \eta \boldsymbol{I} \tag{4.11}$$

where $cov_{\mathbf{p}_j}[\cdot]$ denotes the covariance matrix with respect to the empirical density function $\mathbf{p}_j$. Therefore a global bound for the Lipschitz constant L is given by $\max_{y,j} \|\mathbf{f}_{x_j}(y)\|^2 + \eta \boldsymbol{I}$, which completes the proof. $\qquad\square$

Note that $\mathcal{L}_\eta$ is strongly convex for any positive η. We now present a convergence rate result, using results from (9, Theorem 2.2).

Theorem 4.5. *There exist $\mu, L > 0, \rho > 0$ such that*

$$\|\nabla \mathcal{L}_\eta(\boldsymbol{\theta}_1) - \nabla \mathcal{L}_\eta(\boldsymbol{\theta}_2)\|_{**} \le L \|\boldsymbol{\theta}_2 - \boldsymbol{\theta}_1\|_*$$

$$\mathcal{L}_\eta(\boldsymbol{\theta}_2) \ge \mathcal{L}_\eta(\boldsymbol{\theta}_1) + (\boldsymbol{\theta}_2 - \boldsymbol{\theta}_1)^T \nabla \mathcal{L}_\eta(\boldsymbol{\theta}_1) + \frac{1}{2}\rho\|\boldsymbol{\theta}_2 - \boldsymbol{\theta}_1\|_* \tag{4.12}$$

where $\|\boldsymbol{\theta}\|_* = \sqrt{\boldsymbol{\theta}^T(\boldsymbol{\Sigma}_S^k + \eta\boldsymbol{I})\boldsymbol{\theta}}$ and $\|\boldsymbol{\theta}\|_{**}$ is the corresponding dual norm $\sqrt{\boldsymbol{\theta}^T(\boldsymbol{\Sigma}_S^k + \eta\boldsymbol{I})^{-1}\boldsymbol{\theta}}$*. Furthermore, take $\alpha_k = \frac{1}{L}$ in (4.5), and define $B_k = \|\nabla\mathcal{L}_\eta^k - \boldsymbol{g}_{\mathcal{J}}^k\|^2$, the square error incurred in the gradient at iteration k. Provided a batch growth schedule with $\lim_{k\to\infty}\frac{B_{k+1}}{B_k} \leq 1$, for each iteration (4.5) we have (for any $\epsilon > 0$)*

$$\mathcal{L}_\eta(\boldsymbol{\theta}^k) - \mathcal{L}_\eta(\boldsymbol{\theta}^*) \leq \left(1 - \frac{\rho}{L}\right)^k [\mathcal{L}_\eta(\boldsymbol{\theta}^0) - \mathcal{L}_\eta(\boldsymbol{\theta}^*)] + \mathcal{O}(C_k) , \qquad (4.13)$$

with $C_k = \max\{B_k, (1 - \frac{\rho}{L} + \epsilon)^k\}$.

Proof. Let $\tilde{L}$ denote the bound on the Lipschitz constant of g is provided in (4.10). By the conclusions of Lemma 4.2, we can take $L = \dfrac{1}{\sqrt{\eta + \lambda_{\min}}}\tilde{L}$. Let $\tilde{\rho}$ denote the minimum eigenvalue of (4.11) (note that $\tilde{\rho} \geq \eta$). Then take $\rho = \dfrac{1}{\sqrt{\eta + \lambda_{\max}}}\tilde{\rho}$. The result follows immediately by (9, Theorem 2.2). $\qquad\square$

4.4   Implementation Details

In this section, we briefly discuss important implementation details as well as describe the methods we compare our algorithm with.

4.4.1   Efficient inexact solvers

The linear system we have to invert in iteration (4.5) has very special structure. The matrix $\boldsymbol{\Sigma}$ returned by Algorithm 4.1 may be written as $\boldsymbol{\Sigma} = \boldsymbol{S}\boldsymbol{S}^T$, where each column of $\boldsymbol{S}$ is proportional to one of the vectors $(\mathbf{f}_{x_j}(y) - \mathbf{r})$ computed by the bound. When the dimensions of $\boldsymbol{\theta}$ are large, it is not practical to compute the $\boldsymbol{\Sigma}$ explicitly. Instead, to compute the update in iteration (4.5), we take advantage of the fact that

$$\boldsymbol{\Sigma}\mathbf{x} = \boldsymbol{S}(\boldsymbol{S}^T\mathbf{x}),$$

and use $\boldsymbol{S}$ (computed with a simple modification to the bound method) to implement the action of $\boldsymbol{\Sigma}$. When $\boldsymbol{S} \in \mathbb{R}^{d\times k}$ (k is a mini-batch size), the action of the transpose on a vector can be computed in $O(dk)$, which is very efficient for small k. The action of the regularized curvature approximation $\boldsymbol{\Sigma} + \eta\boldsymbol{I}$ follows immediately. Using only a few iterations of iterative minimization schemes, such as lsqr, conjugate gradient, or others to compute the updates is efficient and further regularizes the subproblems (20, 36).

It is interesting to note that even when $\eta = 0$, and $\boldsymbol{\Sigma}_{\mathcal{J}}$ is not invertible, it makes sense to consider inexact updates. To justify this approach, we present a range lemma. A similar lemma appears in (20) for a different quadratic approximation.

Lemma 4.6. *For any* $\mathfrak{I}$*, we have* $\boldsymbol{\mu}_\mathfrak{I} \in \mathcal{R}(\boldsymbol{\Sigma}_\mathfrak{I})$*.*

Proof. The matrix $\boldsymbol{\Sigma}_\mathfrak{I}$ is formed by a sum of weighted outer products $(\mathbf{f}_{x_j}(y) - \boldsymbol{r})(\mathbf{f}_{x_j}(y) - \boldsymbol{r})^\top$. We can therefore write

$$\boldsymbol{\Sigma}_\mathfrak{I} = \boldsymbol{L}\boldsymbol{D}\boldsymbol{L}^T$$

where $\boldsymbol{L} = [\boldsymbol{l}_1, \ldots, \boldsymbol{l}_{|\Omega|\cdot|\mathfrak{I}|}]$, $\boldsymbol{l}_k = \mathbf{f}_{x_j}(y_k) - \boldsymbol{r}^k$ (k is the current iteration of the bound computation), and $\boldsymbol{D}$ is a diagonal matrix with weights $\boldsymbol{D}_{kk} = \dfrac{1}{|\mathfrak{I}|} \dfrac{\tanh(\frac{1}{2}\log(\alpha_k/z_k))}{2\log(\alpha_k/z_k)}$, where the quantities α_k, z_k correspond to iterations in Algorithm (4.1). Since $\boldsymbol{\mu}$ is in the range of $\boldsymbol{L}$ by construction, it must also be the range of $\boldsymbol{\Sigma}_\mathfrak{I}$. $\qquad\square$

Lemma 4.6 tells us that there is always a solution to the linear system $\boldsymbol{\Sigma}_\mathfrak{I}\Delta\theta = \boldsymbol{\mu}_\mathfrak{I}$, even if $\boldsymbol{\Sigma}_\mathfrak{I}$ is singular. In particular, a minimum norm solution can be found using the Moore-Penrose pseudoinverse, or by simply applying lsqr or cg, which is useful in practice when the dimension d is large. For many problems, using a small number of cg iterations both speeds up the algorithm and serves as additional regularization at the earlier iterations, since the (highly variable) initially small problems are not fully solved.

4.4.2   Mini-batches selection scheme

In our experiments, we use a simple linear interpolation scheme to grow the batch sizes for both the gradient and curvature term approximations. In particular, each batch size (as a function of iteration k) is given by

$$b^k = \min(b^{\text{cap}}, b^1 + \text{round}((k-1)\gamma)),$$

where b^{cap} represents the cap on the maximum allowed size, b^1 is the initial batch size, and γ gives the rate of increase. In order to specify the selections chosen, we will simply give values for each of $(b^1_{\boldsymbol{\mu}}, b^1_{\boldsymbol{\Sigma}}, \gamma_{\boldsymbol{\mu}}, \gamma_{\boldsymbol{\Sigma}})$. For all experiments, the cap $b_{\boldsymbol{\mu}}^{\text{cap}}$ on the gradient computation was the full training set, the cap $b_{\boldsymbol{\Sigma}}^{\text{cap}}$ for the curvature term was taken to be 200, initial $b^1_{\boldsymbol{\mu}}$ and $b^1_{\boldsymbol{\Sigma}}$ were both set to 5. At each iteration of SQB, the parameter vector is updated as follows:

$$\boldsymbol{\theta}_{k+1} = \boldsymbol{\theta}_k - \boldsymbol{\xi}^k,$$

where $\boldsymbol{\xi}^k = \alpha(\boldsymbol{\Sigma}_\mathbb{S}^k + \eta\boldsymbol{I})^{-1}(\boldsymbol{\mu}_\mathfrak{I}^k + \eta\boldsymbol{\theta}^k)$ (α is the step size; we use constant step size for SQB in our experiments). Notice that $\boldsymbol{\xi}^k$ is the solution to the linear system $(\mathbf{\Sigma}_{\mathrm{S}}^k + \eta \mathbf{I})\boldsymbol{\xi}^k = \boldsymbol{\mu}_{\mathrm{J}}^k + \eta \boldsymbol{\theta}^k$ and can be efficiently computed using the lsqr solver or any other iterative solver. For all experiments, we ran a small number of iterations, denoted as l, of lsqr, where l was chosen from the set $\{5, 10, 20\}$, before updating the parameter vector. This technique may be viewed as performing conjugate gradient on the bound. We chose l with the best performance, i.e. the fastest and most stable convergence.

4.4.3   Step size

One of the most significant disadvantages of standard stochastic gradient methods (25, 6) is the choice of the step size. Stochastic gradient algorithms can achieve dramatic convergence rate if the step size is badly tuned (21, 27). An advantage of computing updates using approximated curvature terms is that the inversion also establishes a scale for the problem, and requires minimal tuning. This is well-known in inverse problems. In all experiments below, we used a constant step size; for well conditioned examples we used step size of 1, and otherwise 0.1.

4.4.4   Comparator methods

We compared SQB method with the variety of competitive methods which we list below (the implementation of the last four were obtained from `http://www.di.ens.fr/~mschmidt/Software/SAG.html`):

- L-BFGS: limited-memory BFGS method (quasi-Newton method) tuned for log-linear models which uses both first- and second-order information about the objective function (for L-BFGS this is gradient and approximation to the Hessian); we use the competitive implementation obtained from `http://www.di.ens.fr/~mschmidt/Software/minFunc.html`
- SGD: stochastic gradient descent method with constant step size
- ASGD: averaged stochastic gradient descent method with constant step size
- SAG: stochastic average gradient method using the estimate of Lipschitz constant L_k at iteration k set constant to the global Lipschitz constant
- SAGls: stochastic average gradient method with line search

Since our method uses the constant step size we chose to use the same scheme for the competitor methods like SGD, ASGD and SAG. For those methods we tuned the step size to achieve the best performance (the fastest and most stable convergence). Remaining comparators (L-BFGS and SAGls) use line-search.

4.5 Experiments

We performed experiments with l_2-regularized logistic regression on a binary classification task with regularization parameter $\eta = \frac{1}{T}$. We report the results for six datasets: *rcv1*, *covtype*, and *adult* obtained from `https://www.csie.ntu.edu.tw/~cjlin/libsvmtools/datasets/`, *sido* obtained from `http://www.causality.inf.ethz.ch/home.php`, and *protein* and *quantum* obtained from `http://osmot.cs.cornell.edu/kddcup/`. *rcv1*, *adult*, and *sido* are sparse and the remaining datasets are dense. Table 4.1 summarizes dataset sizes (T and d) and the settings of SQB parameters (l, γ_μ, and γ_{vSigma}). Each dataset was split to training and testing datasets

Table 4.1: Summary of dataset sizes and chosen settings of SQB parameters.

	rcv1	adult	sido	covtype	protein	quantum
T	20242	32561	12678	581012	145751	50000
d	47236	123	4932	54	74	78
l	5	5	5	10	20	5
γ_μ	0.005	0.05	0.01	0.0005	0.005	0.001
γ_Σ	0.0003	0.001	0.0008	0.0003	0.001	0.0008

such that 90% of the original datasets was used for training and the remaining part for testing. Only *sido* and *protein* were split in half to training and testing datasets due to large disproportion of the number of datapoints belonging to each class. The experimental results we obtained are shown in Figures 4.1 and 4.2. We report the training and testing costs as well as the testing error as a function of the number of effective passes through the data and thus the results do not rely on the implementation details. We would like to emphasize however that the average running time for the bound method across the datasets is similar to that of the competitor methods in our experiments. All codes are released and are publicly available at `www.columbia.edu/~aec2163/NonFlash/Papers/Papers.html`.

4.6 Concluding Remarks

We have presented a new semistochastic quadratic bound (SQB) method, together with convergence theory and several numerical examples. The convergence theory is divided into two parts. First, we proved convergence to stationarity of the method under weak hypotheses (in particular, convexity is not required). Second, for the logistic regression problem, we provided a stronger convergence theory, including a rate of convergence analysis.

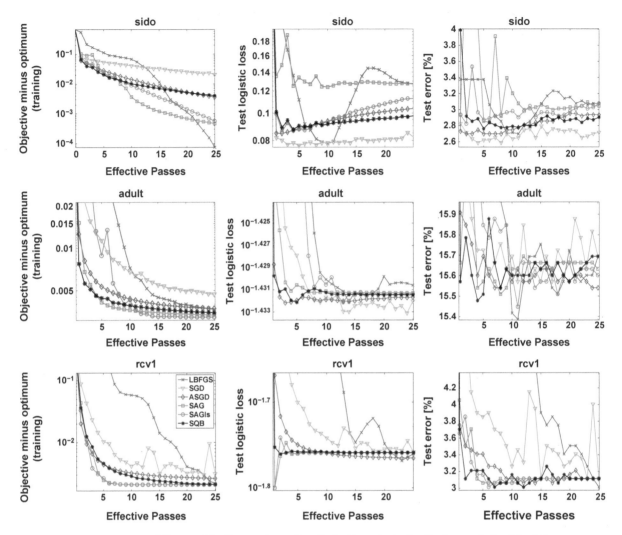

Figure 4.1: Comparison of optimization strategies for l_2-regularized logistic regression. We report training excess cost, testing cost, and testing error. Datasets: *rcv1* ($\alpha_{\text{SGD}} = 10^{-1}$, $\alpha_{\text{ASGD}} = 1$, $\alpha_{\text{SQB}} = 10^{-1}$), *adult* ($\alpha_{\text{SGD}} = 10^{-3}$, $\alpha_{\text{ASGD}} = 10^{-2}$, $\alpha_{\text{SQB}} = 1$), and *sido* ($\alpha_{\text{SGD}} = 10^{-3}$, $\alpha_{\text{ASGD}} = 10^{-2}$, $\alpha_{\text{SQB}} = 1$). *This figure is best viewed in color.*

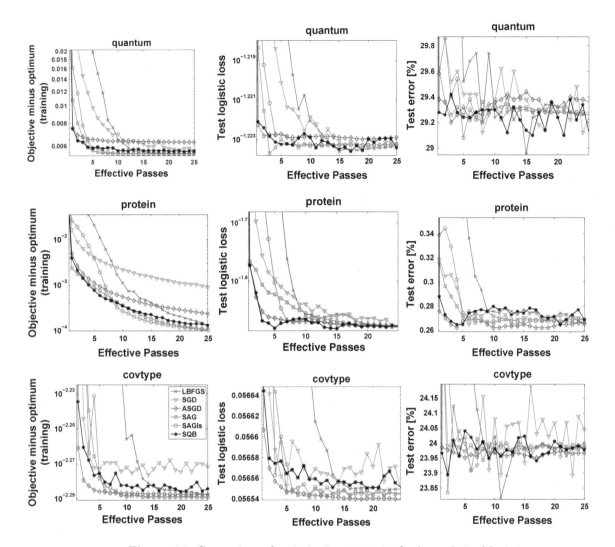

Figure 4.2: Comparison of optimization strategies for l_2-regularized logistic regression. We report training excess cost, testing cost and testing error. Datasets: *covtype* ($\alpha_{\text{SGD}} = 10^{-4}$, $\alpha_{\text{ASGD}} = 10^{-3}$, $\alpha_{\text{SQB}} = 10^{-1}$), *protein* ($\alpha_{\text{SGD}} = 10^{-3}$, $\alpha_{\text{ASGD}} = 10^{-2}$, $\alpha_{\text{SQB}} = 1$) and *quantum* ($\alpha_{\text{SGD}} = 10^{-4}$, $\alpha_{\text{ASGD}} = 10^{-2}$, $\alpha_{\text{SQB}} = 10^{-1}$). *This figure is best viewed in color.*

The main contribution of this chapter is to apply SQB methods in a semi-stochastic large-scale setting. In particular, we developed and analyzed a flexible framework that allows sample-based approximations of the bound from (11) that are appropriate in the large-scale setting, computationally efficient, and competitive with state-of-the-art methods.

Future work includes developing a fully stochastic version of SQB, as well as applying it to learn mixture models and other latent models, as well as models that induce representations, in the context of deep learning.

References

1. A. Agarwal, P. L. Bartlett, P. D. Ravikumar, and M. J. Wainwright. Information-theoretic lower bounds on the oracle complexity of stochastic convex optimization. *IEEE Transactions on Information Theory*, (5):3235–3249.

2. A. Aravkin, M. P. Friedlander, F. Herrmann, and T. van Leeuwen. Robust inversion, dimensionality reduction, and randomized sampling. *Mathematical Programming*, 134(1):101–125, 2012.

3. D. P. Bertsekas and J. N. Tsitsiklis. Gradient convergence in gradient methods with errors. *SIAM J. on Optimization*, 10(3):627–642, July 1999.

4. A. Bordes, L. Bottou, and P. Gallinari. Sgd-qn: Careful quasi-newton stochastic gradient descent. *J. Mach. Learn. Res.*, 10:1737–1754, Dec. 2009.

5. L. Bottou. Online algorithms and stochastic approximations. In D. Saad, editor, *Online Learning and Neural Networks*. Cambridge University Press, Cambridge, UK, 1998.

6. L. Bottou and Y. LeCun. Large scale online learning. In *NIPS*, 2003.

7. R. H. Byrd, G. M. Chin, W. Neveitt, and J. Nocedal. On the use of stochastic hessian information in optimization methods for machine learning. *SIAM Journal on Optimization*, 21(3):977–995, 2011.

8. R. H. Byrd, G. M. Chin, J. Nocedal, and Y. Wu. Sample size selection in optimization methods for machine learning. *Math. Program.*, 134(1):127–155, Aug. 2012.

9. M. P. Friedlander and M. Schmidt. Hybrid deterministic-stochastic methods for data fitting. *SIAM J. Scientific Computing*, 34(3), 2012.

10. M. Hestenes and E. Stiefel. Methods of conjugate gradients for solving linear systems. *Journal of Research of the National Bureau of Standards*, 49(6):409–436, 1952.

11. T. Jebara and A. Choromanska. Majorization for CRFs and latent likelihoods. In *NIPS*, 2012.

12. R. Johnson and T. Zhang. Accelerating stochastic gradient descent using predictive variance reduction. In C. Burges, L. Bottou, M. Welling, Z. Ghahramani, and K. Weinberger, editors, *NIPS*, pages 315–323. 2013.

13. A. J. Kleywegt, A. Shapiro, and T. Homem-de Mello. The sample average approximation method for stochastic discrete optimization. *SIAM J. on Optimization*, 12(2):479–502, Feb. 2002.

14. Q. V. Le, J. Ngiam, A. Coates, A. Lahiri, B. Prochnow, and A. Y. Ng. On optimization methods for deep learning. In *ICML*, 2011.

15. Y. Le Cun, L. Bottou, G. B. Orr, and K.-R. Müller. Efficient backprop. In *Neural Networks, Tricks of the Trade*, Lecture Notes in Computer Science LNCS 1524. Springer Verlag, 1998.

16. N. Littlestone. Learning quickly when irrelevant attributes abound: A new linear-threshold algorithm. *Mach. Learn.*, 2(4):285–318, Apr. 1988.

17. J. Mairal. Optimization with first-order surrogate functions. In *ICML (3)*, pages 783–791, 2013.

18. J. Mairal. Stochastic majorization-minimization algorithms for large-scale optimization. In *NIPS*, pages 2283–2291. 2013.

19. J. Martens. Deep learning via hessian-free optimization. In *ICML*, 2010.

20. J. Martens and I. Sutskever. Training deep and recurrent networks with hessian-free optimization. In G. Montavon, G. Orr, and K.-R. Mller, editors, *Neural Networks: Tricks of the Trade*, volume 7700 of *Lecture Notes in Computer Science*, pages 479–535. Springer Berlin Heidelberg, 2012.

21. A. Nemirovski, A. Juditsky, G. Lan, and A. Shapiro. Robust stochastic approximation approach to stochastic programming. *SIAM J. on Optimization*, 19(4):1574–1609, Jan. 2009.

22. Y. Nesterov. *Introductory lectures on convex optimization : a basic course.* Applied optimization. Kluwer Academic Publ., Boston, Dordrecht, London, 2004.

23. Y. Nesterov. Primal-dual subgradient methods for convex problems. *Math. Program.*, 120(1):221–259, 2009.

24. B. T. Polyak and A. B. Juditsky. Acceleration of stochastic approximation by averaging. *SIAM J. Control Optim.*, 30(4):838–855, July 1992.

25. H. Robbins and S. Monro. A Stochastic Approximation Method. *The Annals of Mathematical Statistics*, 22(3):400–407, 1951.

26. F. Rosenblatt. The perceptron: A probabilistic model for information

storage and organization in the brain. *Psychological Review*, 65(6):386–408, 1958.

27. N. L. Roux, M. W. Schmidt, and F. Bach. A stochastic gradient method with an exponential convergence rate for finite training sets. In *NIPS*, 2012.

28. N. N. Schraudolph. Local gain adaptation in stochastic gradient descent. In *ICANN*, 1999.

29. N. N. Schraudolph. Fast curvature matrix-vector products for second-order gradient descent. *Neural Computation*, 14:2002, 2002.

30. N. N. Schraudolph, J. Yu, and S. Günter. A stochastic quasi-newton method for online convex optimization. In *AISTATS*, 2007.

31. S. Shalev-Shwartz and T. Zhang. Proximal stochastic dual coordinate ascent. *CoRR*, abs/1211.2717, 2012.

32. S. Shalev-Shwartz and T. Zhang. Stochastic dual coordinate ascent methods for regularized loss minimization. *CoRR*, abs/1209.1873, 2012.

33. S. Shalev-Shwartz and T. Zhang. Accelerated mini-batch stochastic dual coordinate ascent. *CoRR*, abs/1305.2581, 2013.

34. A. Shapiro and T. H. de Mello. On the rate of convergence of optimal solutions of monte carlo approximations of stochastic programs. *SIAM Journal on Optimization*, 11:70–86, 2000.

35. P. Tseng. An incremental gradient(-projection) method with momentum term and adaptive stepsize rule. *SIAM J. on Optimization*, 8(2):506–531, Feb. 1998.

36. C. R. Vogel. *Computational Methods for Inverse Problems*. Society for Industrial and Applied Mathematics, 2002.

37. C. Wang, X. Chen, A. Smola, and E. Xing. Variance reduction for stochastic gradient optimization. In *NIPS*, pages 181–189. 2013.

38. A. S. Y. Wardi. Convergence analysis of stochastic algorithms. *Mathematics of Operations Research*, 21(3):615–628, 1996.

5 Use of Deep Learning Features in Log-Linear Models

Li Deng   deng@microsoft.com
Microsoft Research
Redmond, WA, USA

A log-linear model, popular for discriminative sequence modeling and classification, by itself is a shallow architecture given fixed, non-adaptive, human-engineered feature functions. However, the flexibility of the log-linear model in using many kinds of feature functions as its input allows the exploitation of diverse high-level features computed automatically from deep learning systems. We propose and explore a paradigm of connecting the deep leaning features as inputs to log-linear models, which, in combination with the feature hierarchy, form a powerful deep sequence classifier. Three case studies are provided in this chapter to instantiate this paradigm. First, deep stacking networks and their kernel versions are used to provide deep learning features for a static log-linear model—the softmax classifier or maximum entropy model. Second, deep-neural-network features are extracted to feed to a sequential log-linear model—the conditional random field. And third, a log-linear model is used as a stacking-based ensemble learning machine to integrate a number of deep learning systems' outputs. All these three types of deep classifier have their effectiveness verified in experiments. Finally, compared with the traditional log-linear modeling approach which relies on human feature engineering, we point out one main weakness of the new framework in its lack of ability to naturally embed domain knowledge. Future directions are discussed for overcoming this weakness by integrating deep neural networks with deep generative models.

5.1   Introduction

Log-linear modeling forms the basis of a class of important machine learning methods that have found wide applications, notably in human language technology including speech and natural language processing (53, 58, 41, 49, 38, 39). In mathematical terms, a log-linear model takes the form of a function whose logarithm is a linear function of the model parameters:

$$Ce^{\sum_i w_i f_i(\mathbf{x})} \quad or \quad Ce^{\mathbf{w}^T \boldsymbol{f}(\mathbf{x})} \tag{5.1}$$

where $f_i(\mathbf{x})$ are functions of the input data variables $\mathbf{x}$, in general a vector of values, and w_i are model parameters (the bias term can be absorbed by introducing an additional "feature" of $f_i(\mathbf{x}) = 1$). Here, C does not depend on the model parameters w_i but may be a function of data $\mathbf{x}$.

A special form of the log-linear model of Eq.(5.1) is the softmax function, which has the following form that is often used to model the class-posterior distribution for classification problems with multiple ($K > 2$) classes:

$$P(y = j|\mathbf{x}) = \frac{e^{s_j}}{\sum_{k=1}^{K} e^{s_k}}, \quad where \quad s_j = \boldsymbol{w}_j^T \mathbf{x}, \quad j = 1, 2, ..., K \tag{5.2}$$

The reason for calling this function "softmax" is that the effect of exponentiating the K values of $s_1, s_2, \ldots, s_K$ in the exponents of Eq.(5.2) is to exaggerate the differences between them. As a result, the softmax function will return a value close to zerqo whenever s_j is significantly less than the maximum of all the values, and will return a value close to one when applied to the maximum value, unless it is extremely close to the next-largest value. Therefore, softmax can be used to construct a weighted average that behaves as a smooth function to approximate the non-smooth function $\max(s_1, s_2, \ldots, s_K)$.

The classifiers exploiting the softmax function of Eq.(5.2) are often called softmax regression (or classifiers), multinomial (or multiclass) logistic regression, multinomial regression (or logit), maximum entropy (MaxEnt) classifiers, or conditional maximum entropy models. These form a class of popular classification methods in statistics, machine learning, and in speech and language processing that generalize logistic regression from two-class to multi-class problems. Such classifiers can be used to predict the probabilities of the different possible outcomes of a categorically distributed dependent variable, given a set of independent variables which may be real-valued, binary-valued, categorical-valued, etc. Extending the softmax classifier for static patterns to sequential patterns, we have the conditional random field (CRF) as a more general case of log-linear classifiers (53, 58, 79).

Log-linear models are interesting from several points of view. First, many popular generative models (Gaussian models, word-counting models, part-of-speech tagging, etc.) in speech and language processing have the posterior form that is shown to be equivalent to log-linear models (41). This provides important insight into the connections between generative and discriminative learning paradigms which have been commonly regarded as two separate classes of approaches (18). Second, the natural training criterion of log-linear models is the logarithm of the class posterior probability or conditional maximum likelihood. This gives rise to a convex optimization problem and the margin concept can be built into the training. Third, log-linear models can be extended to include hidden variables, thereby connecting naturally to the common generative Gaussian mixture and hidden Markov models (36, 80).

Perhaps most importantly, the softmax classifier as an instance of the log-linear model has been used extensively in recent years as the top layer in deep neural networks (DNNs) that consist of many other (lower) layers as well (81, 13, 71, 63, 69, 82, 62, 61, 31, 32). The CRF sequence classifier has also been used in a similar fashion, i.e., being connected to a DNN whose parameters are tied across the entire sequence (60, 51). The conventional way of interpreting the sweeping success of the DNN in speech recognition, from small to large tasks and from laboratory experiments to industrial deployments, is that the DNN is learned discriminatively in an end-to-end manner (72, 45, 83, 23). The excellent scalability of the DNN and its huge capacity provided by massive weight parameters and distributed representations (21) have enabled its success. In this chapter, we provide a new way of looking at the DNN and other deep learning models in terms of their ability to provide high-level features to relatively simple classifiers such as log-linear models, rather than viewing an entire deep learning model as a complex classifier. That is, we connect deep learning to log-linear classifiers via the separation of feature extraction and classification.

In the early days of speech recognition research when shallow models such as Gaussian mixture model (GMM) and hidden Markov model (HMM) were exploited (66, 3, 4), integrated learning of classifiers and feature extractors was shown to outperform that when the two stages are separated (6, 10, 11). Some earlier successes of the DNN in speech recognition also adopted the integrated or end-to-end learning via backpropagating errors all the way from top to the bottom in the network (59). However, more recent successes of the DNN and more advanced deep models have shown that using the deep models to provide features for separate classifiers has numerous advantages over integrated learning (75, 78, 17). For example, this feature-based approach enables the use of existing GMM-HMM discriminative training methods and infrastructure developed and matured over many years

(78), and it helps transfer or multitask deep learning (35, 42, 46). This type of large-scale discriminative training, unlike end-to-end training of the DNN by mini-batch-based backpropagation, is naturally suited for batch-based parallel computation since it rests on extended Baum-Welch algorithm (65, 40, 77). Also, speaker-adaptive (1) and noise-adaptive training techniques (26, 50, 34) successfully developed for the GMM-HMM can be usefully applied. Further, making use of the features derived from the DNN, we can easily perform multi-task and transfer learning. Successful applications of this type have been shown in multilingual and mixed-bandwidth speech recognition (42, 46), which had much less success in the past using the GMM-HMM approach (57). Another benefit of using deep learning features is that it avoids overfitting, especially when the depth of the network grows very large as in the deep stacking network (DSN) to be discussed in a later section. Moreover, when feature extraction performed by deep models is separated from the classification stage in the overall system, different types of deep learning methods can be more easily combined (9, 20) and unsupervised learning methods (e.g., autoencoders) may be more naturally incorporated (28, 68, 54).

This chapter is aimed to explore the topic of log-linear modeling as supervised classifiers using features automatically derived from deep learning systems. Specifically, we focus on the use of log-linear models as the classifier, whose flexibility for accepting a wide range of features facilitates the use of deep learning features. In Section 2, a framework of using deep learning features for performing supervised learning tasks is developed with examples given in the context of acoustic modeling for speech recognition. Sections 3-5 provide three more detailed case studies on how this general framework is applied. The case study presented in Section 3 concerns the use a specialized deep learning architecture, the DSN and its kernel version, to compute the features for static log-linear or max-entropy classifiers. Some key implementation details and experimental results are included, not published in the previous literature. Section 4 shows how the standard DNN features can be interfaced to a dynamic log-linear model, the conditional random field, whose standard learning leads to the equivalent full-sequence learning for the DNN-HMM which gives the state of the art accuracy in large vocabulary speech recognition as of the writing of this chapter. The final case study reported in Section 5 makes use of the log-linear model as a stacking mechanism to perform ensemble learning, where three deep learning systems (deep forward and fully-connected neural network, deep convolutional neural network, and recurrent neural network) provide three separate streams of deep learning features for system combination in a log-linear fashion.

5.2   A Framework of Using Deep Learning Features for Classification

Deep Learning is a class of machine learning techniques, where many layers of information processing stages in hierarchical architectures are exploited. The most prominent successes of deep learning, achieved in recent years, are in supervised learning for classification tasks (45, 14, 52). The essence of deep learning is to compute hierarchical features or representations of the observational data, where the higher-level features or factors are defined from lower-level ones. Recent overviews of deep learning can be found in (5, 70, 23).

Given fixed feature functions $f(\mathbf{x})$, the log-linear model defined in Eq.(5.1) is a shallow architecture. However, the flexibility of the log-linear model in using the feature functions (i.e., no restriction on the form of functions in $f(\mathbf{x})$) allows the model to exploit a wide range of high-level features including those computed from separate deep learning systems. In this section, we explore a general framework of connecting such deep leaning features as the input to the log-linear model. A combination of deep learning systems (e.g., the DNN) and the log-linear model gives rise to powerful deep architectures for classification tasks.

To build up this unifying framework, let us start with the (shallow) architecture of the GMM-HMM and put the discussion in the specific context of acoustic modeling for speech recognition. Before around 2010-2011, speech recognition technology had been dominated by the HMM, where each state is characterized by the GMM; .hence, the GMM-HMM (66, 24). In Figure 5.1, we illustrate the GMM-HMM system by showing how the raw speech feature vectors indexed by time frame t, $\mathbf{x}_t$, such as Mel-Frequency Cepstral Coefficients (MFCCs) and Perceptual Linear Prediction (PLP) features, form a feature sequence to feed into the GMM-HMM as a sequence classifier, producing a sequence of linguistic symbols (e.g., words, phones, etc.) as the recognizer's output.

While significant technological successes had been achieved using complex and carefully engineered variants of GMM-HMMs and acoustic features suitable for them, researchers long before that time had clearly realized that the next generation of speech recognition technology would require solutions to many new technical challenges under diversified deployment environments and that overcoming these challenges would likely require "deep" architectures that can at least functionally emulate the human speech system known to have dynamic and hierarchical structure in both production and perception (73, 33, 19, 25, 8, 64). An attempt to incorporate a primitive level of understanding of this deep speech structure, initiated at the 2009

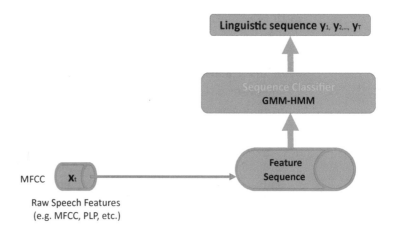

Figure 5.1: Illustration of the GMM-HMM-based speech recognition system: Feeding low-level speech sequences into the GMM-HMM sequence classifier.

NIPS Workshop on Deep Learning for Speech Recognition and Related Applications (27), has helped ignite the interest of the speech community in pursuing a deep representation learning approach based on the deep neural network (DNN) architecture. The generative pre-training method in effective learning of the DNN was pioneered by the machine learning community a few years earlier (43, 44). The DNN, learned both generatively or in a purely discriminative manner when large training data and computing resources are available, has rapidly evolved into the new state of the art in speech recognition with pervasive industry-wide adoption (14, 45, 23).

The DNN by itself is a static classifier and does not handle variable-dimensional sequences as in the raw speech data. The DNN is shown in the left portion of Figure 5.2, where $\mathbf{h}_{1,t}, \mathbf{h}_{2,t}$, and $\mathbf{h}_{3,t}$ illustrate three hidden state vectors of the DNN for the low-level speech feature vector $\mathbf{x}_t$ at time t, and $\boldsymbol{y}_t$ is the corresponding output vector of the DNN. Denoted by $\boldsymbol{d}_t$ is the desired sequence of target vectors (often coded as sparse one-hot vectors) used for training the DNN using either the cross-entropy (CE) or maximum-mutual information (MMI) criteria. Now turn to the right portion of Figure 5.2. The high-level DNN features extracted from the $\hat{v}_{3,t}$ layer is shown to feed into the sequence classifier (a log-linear model followed by an HMM), which produces symbolic linguistic sequences. The overall architecture shown in Figure 5.2 using DNN-derived features for a log-linear model followed by an HMM is called the DNN-HMM. The parameters of the overall DNN-HMM system can be learned either using backpropagation with the end-to-end style where the errors defined at the sequence classifier are propagated back all the way into all the hidden layers of the DNN, or propagating the errors into only the high-level feature level with the DNN parameters learned separately using the the error criterion defined at the DNN's output layer as shown in the left portion of Figure 5.2. The former, end-to-end learning is likely to work better with large amounts of training data. The latter, based on fixed and high-level features, is less prone to overfitting, and is also more effective in multi-task learning since the high-level features tend to transfer well from one learning task to another as demonstrated in multi-lingual speech recognition as we discussed in Section 1.

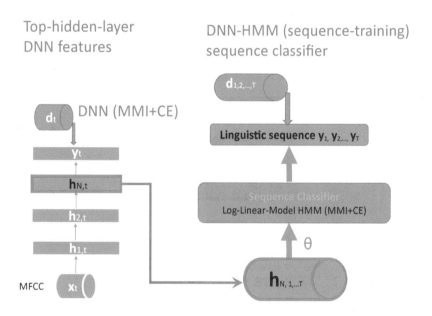

Figure 5.2: Illustration of the DNN-HMM-based speech recognition system: Feeding high-level DNN-derived feature sequences into a log-linear sequence classifier.

While the DNN-HMM architecture shown in Figure 5.2 produces much lower errors than the previous state-of-the-art GMM-HMM systems, it is only one of many possible deep architectures. For example, not just one hidden state vector in the DNN can serve as the high-level features for the log-linear sequence classifier, a combination of them, typically in a straightforward form of concatenation, can serve as more powerful DNN features as demonstrated in (17), which is shown in the left portion of Figure 5.3. Further, the sequence classifier may not be limited to the log-linear model. Other sequence classifiers can take the high-level DNN features as their inputs also, which is shown in the right portion of Figure 5.3. Specifically, the use of GMM-HMM sequence classifiers for DNN-derived features is shown to almost match the low error rate produced by the DNN-HMM system (78).

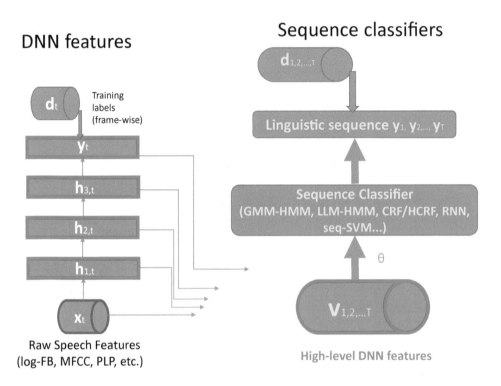

Figure 5.3: Extensions of the system of Figure 5.2 in two ways. First, various hidden layers and the output layer of the DNN can be combined (e.g., via concatenation) to form the high-level DNN features. Second, the sequence classifier can be extended to many different types.

In summary, in this section we present a general framework of using deep learning features for sequence classification, with the application examples drawn from speech recognition. This framework enables us to naturally connect the DNN features as the input to the log-linear model as the most prominent scheme for sequence classification. Importantly, as a special case of this framework, where the DNN feature is derived from the top hidden layer of the DNN and where the sequence classifier uses the softmax followed by an HMM, we recover the popular DNN-HMM architecture widely used in current state-of-the-art speech recognition systems. In the next three sections, we will provide three special cases of this framework, some aspects of which have been published in recent literature but can be better understood in a unified manner based on the framework discussed in this section.

5.3 DSN and Kernel-DSN Features for Log-Linear Classifiers

In this section, we discuss the use of deep learning features computed from the Deep stacking networks (DSN) and its kernel version (K-DSN) for log-linear models. We first review the DSN and K-DSN and study their properties, and then exploit them as feature functions to connect to the log-linear models, where application case studies are provided.

5.3.1 Deep stacking networks (DSN)

Stacking is a class of machine learning techniques that form combinations of different predictors' outputs in order to give improved overall prediction accuracy, typically through improved generalization (76, 7). In (12), for example, stacking is applied to reduce the overall generalization error of a sequence of trained predictors by generating an unbiased set of previous predictions' outputs for use in training each successive predictor.

This concept of stacking is more recently applied to construct a deep network where the output of a (shallow) network predictor module is used, in conjunction with the original input data, to serve as the new, expanded "inputs" for the next level of the network predictor module in the full, multiple-module network, which is called the Deep Stacking Network (DSN) (22, 30, 47). Figure 5.4 shows a DSN with three modules, each with a different color and each consisting of three layers with upper-layer weight matrix denoted by $\mathbf{U}$ and lower-layer weight matrix denoted by $\mathbf{W}$.

In the DSN, different modules are constructed somewhat differently. The lowest module comprises the following three layers. First, there is a linear layer with a set of input units. They correspond to the raw input data in the vectored form. Let N input vectors in the full training data be $\mathbf{X} = [\mathbf{x}_1, ..., \mathbf{x}_i, ..., \mathbf{x}_N]$, with each vector $\mathbf{x}_i = [\mathbf{x}_{1i}, ..., \mathbf{x}_{ji}, ..., \mathbf{x}_{Di}]^T$. Then, the input units correspond to the elements of $\mathbf{x}_i$, with dimensionality D. Second, the non-linear layer consists of a set of sigmoidal hidden units. Denote by L the number of hidden units and define

$$\mathbf{h}_i = \sigma(\mathbf{W}^T \mathbf{x}_i) \tag{5.3}$$

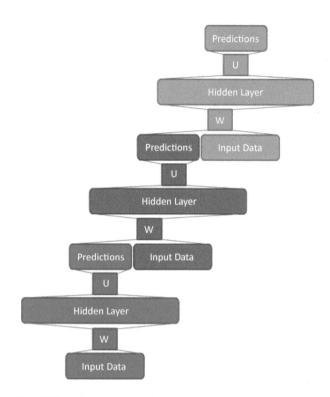

Figure 5.4: Illustration of a DSN with three modules, each with a different color. Each module consists of three layers connected by upper weight matrix denoted by **U** and lower weight matrix denoted by **W**.

as the hidden layer's output, where $\sigma(.)$ is the sigmoid function and **W** is an $D \times L$ trainable weight matrix, at the bottom module, acting on the input layer. Note the bias vector is implicitly represented in the above formulation when $\mathbf{x}_i$ is augmented with all ones. Third, the output layer consists of a set of C linear output units with their values computed by $\mathbf{y}_i = \mathbf{U}^T \mathbf{h}_i$, where **U** is an $L \times C$ trainable weight matrix associated with the upper layer of the bottom module. Again, we augment $\mathbf{h}_i$ with a vector consisting of all one's. The output units represent the targets of classification (or regression).

Above the bottom one, all other modules of a DSN, which are stacking up one above another, are constructed in a similar way to the above but with a key exception in the input layer. Rather than making the input units take the raw data vector, we concatenate the raw data vector with the output layer(s) in the lower module(s). Such an augmented vector serves as the "effective input" to the immediately higher module. The dimensionality, D_m, of the augmented input vector is a function of the module number, m, counted from bottom up according to

$$D_m = D + C(m - 1), \quad m = 1, 2, \cdots, M \tag{5.4}$$

where $m = 1$ corresponds to the bottom module.

A closely related difference between the bottom module and the remaining modules concerns the augmented weight matrix $\mathbf{W}$. The weight matrix augmentation results from the augmentation of the input units. That is, the dimensionality of $\mathbf{W}$ changes from $D \times L$ to $D_m \times L$. Additional columns of the weight matrix corresponding to the new output units from the lower module(s) are initialized with random numbers, which are subject to optimization.

For each module, given $\mathbf{W}$, learning $\mathbf{U}$ is a convex optimization problem. The solution differs for separate modules mainly in ways of setting the lower-layer weight matrices $\mathbf{W}$ in each module, which varies its dimensionality across modules according to Eq. 5.4, before applying the learning technique presented below.

In the supervised learning setting, both training data $\mathbf{X} = [\mathbf{x}_1, ..., \mathbf{x}_i, ..., \mathbf{x}_N]$ and the corresponding labeled target vectors $\mathbf{T} = [\mathbf{t}_1, ..., \mathbf{t}_i, ..., \mathbf{t}_N]$, where each target $\mathbf{t}_i = [\mathbf{t}_{1i}, ..., \mathbf{t}_{ji}, ..., \mathbf{t}_{Ci}]^T$, are available. We use the loss function of mean square error to learn weight matrices $\mathbf{U}$ assuming $\mathbf{W}$ is given. That is, we aim to minimize:

$$E = Tr[(\mathbf{Y} - \mathbf{T})(\mathbf{Y} - \mathbf{T})^T], \tag{5.5}$$

where $\mathbf{Y} = [\mathbf{y}_1, ..., \mathbf{y}_i, ..., \mathbf{y}_N]$.

Importantly, if the weight matrix $\mathbf{W}$ is determined already (e.g., via judicious initialization), then the hidden layer values $\mathbf{H} = [\mathbf{h}_1, ..., \mathbf{h}_i, ..., \mathbf{h}_N]$ are also determined. Consequently, the upper-layer weight matrix $\mathbf{U}$ in each module can be determined by setting the gradient

$$\frac{\partial E}{\partial \mathbf{U}} = 2\mathbf{H}(\mathbf{U}^T\mathbf{H} - \mathbf{T})^T \tag{5.6}$$

to zero. This is a well established convex optimization problem and has a straightforward closed-form solution, known as the pseudo-inverse:

$$\mathbf{U} = (\mathbf{H}\mathbf{H}^T)^{-1}\mathbf{H}\mathbf{T}^T. \tag{5.7}$$

Combining Eqns. (5.3) and (5.7), we see that $\mathbf{U}$ is an explicit function of $\mathbf{W}$, denoted, say, as

$$\mathbf{U} = F(\mathbf{W}). \tag{5.8}$$

Note that Eq. (5.8) provides a powerful constraint when learning matrix $\mathbf{W}$; i.e.

$$\hat{\mathbf{W}} = \underset{\mathbf{W}}{\operatorname{argmax}} E(\mathbf{U}^*, \mathbf{W}), \quad subject\ to\ \mathbf{U}^* = F(\mathbf{W}). \tag{5.9}$$

When gradient descent method is used to optimize $\mathbf{W}$, we seek to compute the *total* derivative of the error function:

$$\frac{dE}{d\mathbf{W}} = \frac{\partial E}{\partial \mathbf{W}} + \frac{\partial E}{\partial \mathbf{U}^*}\frac{\partial \mathbf{U}^*}{\partial \mathbf{W}}. \tag{5.10}$$

This total derivative can be found in a direct analytical form (i.e., without recursion as in backpropagation). An easy way to pursue is to remove the constraint in Eq. (5.9) by substituting the constraint directly into the objective function, yielding the unconstrained optimization problem of

$$\hat{\mathbf{W}} = \underset{\mathbf{W}}{\operatorname{argmax}}\, E[F(\mathbf{W}), \mathbf{W}]. \tag{5.11}$$

The analytical form of the total derivative is derived below:

$$
\begin{aligned}
\frac{dE}{d\mathbf{W}} &= \frac{dTr[(\mathbf{U}^T\mathbf{H} - \mathbf{T})(\mathbf{U}^T\mathbf{H} - \mathbf{T})^T]}{d\mathbf{W}} \\[2mm]
&= \frac{dTr[([(\mathbf{HH}^T)^{-1}\mathbf{HT}^T]^T\mathbf{H} - \mathbf{T})([(\mathbf{HH}^T)^{-1}\mathbf{HT}^T]^T\mathbf{H} - \mathbf{T})^T]}{d\mathbf{W}} \\[2mm]
&= \frac{dTr[\mathbf{TT^T} - \mathbf{TH^T}(\mathbf{HH^T})^{-1}\mathbf{HT^T}]}{d\mathbf{W}} \\[2mm]
&= \frac{dTr[(\mathbf{HH}^T)^{-1}\mathbf{HT}^T\mathbf{TH}^T]}{d\mathbf{W}} \\[2mm]
&= \frac{dTr[(\sigma(\mathbf{W^T X})[\sigma(\mathbf{W^T X})]^\mathbf{T})^{-1}\sigma(\mathbf{W^T X})\mathbf{T^T T}[\sigma(\mathbf{W^T X})]^\mathbf{T}]}{d\mathbf{W}} \\[2mm]
&= 2\mathbf{X}[\mathbf{H}^T \circ (1 - \mathbf{H})^T \circ [\mathbf{H}^\dagger(\mathbf{HT}^T)(\mathbf{TH}^\dagger) - \mathbf{T}^T(\mathbf{TH}^\dagger)] \qquad (5.12)
\end{aligned}
$$

where $\mathbf{H}^\dagger = \mathbf{H}^T(\mathbf{HH}^T)^{-1}$ and $\circ$ denotes element-wise matrix multiplication. In deriving Eq.5.12, we used the fact that $\mathbf{HH}^T$ is symmetric and so is $(\mathbf{HH}^T)^{-1}$.

Importantly, the total derivative in Eq.5.12 with respect to lower-layer weight matrix $\mathbf{W}$ is different from the gradient computed in the standard backpropagation algorithm which requires recursion through the hidden layer and which is partial derivative with respective to $\mathbf{W}$ instead of the total derivative. That is, the backpropagation algorithm gives only the first term $\frac{\partial E}{\partial \mathbf{W}}$ in Eq.5.10. Thus the difference between the gradients used in the backpropagation algorithm and in the algorithm based on Eq.5.12 differs by the quantity of

$$\frac{\partial E}{\partial \mathbf{U}^*} \frac{\partial \mathbf{U}^*}{\partial \mathbf{W}}, \tag{5.13}$$

where $\mathbf{U}^* = F(\mathbf{W}) = (\mathbf{H}\mathbf{H}^T)^{-1}\mathbf{H}\mathbf{T}^T$ and each vector component of matrix $\mathbf{H}$ is $\mathbf{h}_i = \sigma(\mathbf{W}^T \mathbf{x}_i)$.

This difference may account for why learning the DSN using batch-mode gradient descent using the total derivative of Eq.5.12 is more effective than using batch-mode backpropagation based on partial derivative experimentally (22). It is noted that using the total derivative of Eq.5.12 is equivalent to alternating minimization algorithm with an "infinite" step size to achieve the global optimum along the "coordinate" of updating $\mathbf{U}$ while fixing $\mathbf{W}$.

Applications of the DSN features discussed in this section for log-linear classifiers will be presented in Section 4.

5.3.2 Kernel deep stacking networks (K-DSN)

It is well known that mapping raw speech, audio, text, image, and video data into desirable feature vectors for machine learning algorithms to consume typically require extensive human expertise, intuition, and domain knowledge. Kernel methods are a powerful set of tools that alleviate such difficult requirements via the kernel "trick". Deep learning methods, on the other hand, avoid hand-designed resources and time-intensive feature engineering by adopting layer-by-layer generative or discriminative learning. Understanding the essence of these two styles of feature mapping and their connections is of both conceptual and practical significance. First, kernel methods can be naturally extended to overcome the linearity inherent in the pattern predictive functions. Second, deep learning methods can also be enhanced to let the effective hidden feature dimensionality grow to infinity without encountering otherwise computational and overfitting difficulties. Both aspects of the extension lead to integrated kernel and deep learning architectures, which perform better in practical applications. One such architecture, which integrates the kernel trick in the DSN and is called Kernel-DSN or K-DSN (29), is discussed in this section. Specifically, the insights into how the generally finite-dimensional (hidden) features in the DSN can be transformed into infinite-dimensional features via the kernel trick without incurring computational and regularization difficulty are elaborated.

The DSN architecture discussed in the preceding section has convex learning for the weight matrix $\mathbf{U}$ given the hidden layers' outputs in each module, but the learning of the weight matrix $\mathbf{W}$ is non-convex. For most applications, the size of $\mathbf{U}$ is comparable to that of $\mathbf{W}$ and then the DSN is not strictly a convex network. In a recent extension of the DSN, a tensor structure was imposed, shifting the majority of non-convex learning burden for $\mathbf{W}$ into a convex one (47, 48). In the K-DSN extension discussed here, non-convex learning for $\mathbf{W}$ is completely eliminated using the kernel trick. In deriving the K-DSN architecture and the associated learning algorithm, the sigmoidal hidden layer $\mathbf{h}_i = \sigma(\mathbf{W}^T \mathbf{x}_i)$ in a DSN module is generalized into a generic nonlinear mapping function $\mathbf{G}(\mathbf{X})$ from raw input features $\mathbf{X}$. The high or possibly infinite dimensionality in $\mathbf{G}(\mathbf{X})$ is determined only implicitly by the kernel function.

It can be derived that for each new input vector $\mathbf{x}$ in the test set, a (bottom) module of the K-DSN has the prediction function of

$$\mathbf{y}(\mathbf{x}) = \mathbf{k}^T(\mathbf{x})(C\mathbf{I} + \mathbf{K})^{-1}\mathbf{T} \tag{5.14}$$

where $\mathbf{T} = [\mathbf{t}_1, ..., \mathbf{t}_i, ..., \mathbf{t}_N]$ are the target vectors for training, and the kernel vector $\mathbf{k}(\mathbf{x})$ is defined such that its elements have values of $k_n(\mathbf{x}) = k(\mathbf{x}_n, \mathbf{x})$ in which $\mathbf{x}_n$ is a training sample and $\mathbf{x}$ is the current test sample. For the non-bottom DSN module of $l \geq 2$, the same prediction function hold except the kernel matrix is expanded to

$$\mathbf{K}^{(l)} = \mathbf{G}\left([\mathbf{X}|\mathbf{Y}^{(l-1)}|\mathbf{Y}^{(l-2)}|...\mathbf{Y}^{(1)}]\right)\mathbf{G}^T\left([\mathbf{X}|\mathbf{Y}^{(l-1)}|\mathbf{Y}^{(l-2)}|...\mathbf{Y}^{(1)}]\right) \tag{5.15}$$

Comparing the prediction in the DSN and that of the K-DSN, we can identify key advantages of the K-DSN. Unlike the DSN which need to compute hidden units' outputs, the K-DSN does not need to explicitly compute hidden units' outputs. Let's take the example of Gaussian kernel, where the kernel trick gives the equivalent of an infinite number of hidden units without the need to compute them explicitly. Further, one no longer needs to learn the lower-layer weight matrix $\mathbf{W}$ as in the DSN, and the kernel parameter (e.g., the single variance parameter σ in the Gaussian kernel) makes the K-DSN much less subject to overfitting.

The architecture of a K-DSN using the Gaussian kernel is shown in Figure 5.5, where the entire network is characterized by two module (l)-dependent hyper-parameters ($l = 1, 2, ..., L$):

- $\sigma^{(l)}$, which is the kernel smoothing parameter, and

- $C^{(l)}$, which is the regularization parameter.

While both of these parameters are intuitive and their tuning (via linesearch or leave-one-out cross validation) is straightforward for a sin-

gle bottom module, tuning them from module to module is more difficult.
It was found in experiments that if the bottom module is tuned too well,
then adding more modules would not benefit much. In contrast, when the
lower modules are loosely tuned (i.e., relaxed from the results obtained from
straightforward methods), the overall K-DSN often performs much better.
In practice, these hyperparameters can be determined using empirical tuning schedules (e.g., RPROP-like methods (67)) to adaptively regularize the
K-DSN from bottom to top modules.

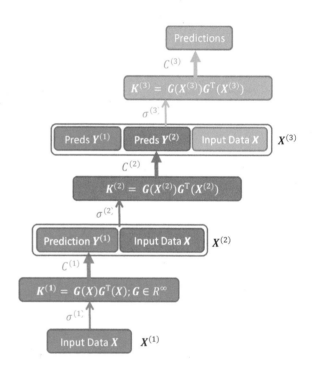

Figure 5.5: The architecture of a K-DSN using Gaussian kernel.

5.3.3   Connecting DSN and K-DSN features to classification models

Based on the general framework of using deep learning features for classification as presented in Section 2, we now describe in this section selected
issues of importance and related experiments on creating features out of the
DSN and K-DSN discussed earlier for a separate classifier (e.g., a log-linear
model) for classification tasks.

Some earlier experimental results using DSNs for MNIST digit recognition (with no elastic distortion) and TIMIT phone recognition tasks were published in (30). Here some key aspects of the DSN design are summarized including two unpublished aspects of the design and also including adding a softmax classifier on top of the DSN. Related new experimental results are then reported in this section.

Given the basic DSN design described earlier in this section, an important aspect is how to initialize the weight matrix in each of the DSN module. It was reported in (30) that initialization using a standard RBM (i.e., with symmetric up- and down-weights) gives a significant accuracy gain over random initialization. If we break this symmetry by using asymmetric weights in the RBM, where the up-weights are constrained to be a fixed factor of the down-weights, then better results can be obtained. The scaled factor is determined by the square root of the ratio of the number of RBM hidden units over that of visible units. The motivation is that with different numbers of units in the hidden and visible layers, the equal-weight constraint in the RBM would not balance the average activities of the two layers.

We can further improve classification accuracy of a DSN by regularizing its learning by incorporating reconstruction errors in the gradient computation. That is, we replace the gradient computation in Eq.(5.12):

$$\frac{dE}{d\mathbf{W}} = 2\mathbf{X}[\mathbf{H}^T \circ (1 - \mathbf{H})^T \circ [\mathbf{H}^\dagger(\mathbf{H}\mathbf{T}^T)(\mathbf{T}\mathbf{H}^\dagger) - \mathbf{T}^T(\mathbf{T}\mathbf{H}^\dagger)] \qquad (5.16)$$

by

$$\begin{aligned}
\frac{dE}{d\mathbf{W}} =\ & 2\mathbf{X}[\mathbf{H}^T \circ (1 - \mathbf{H})^T \circ [\mathbf{H}^\dagger(\mathbf{H}\mathbf{T}^T)(\mathbf{T}\mathbf{H}^\dagger) - \mathbf{T}^T(\mathbf{T}\mathbf{H}^\dagger)] \quad (5.17) \\
& + \ \gamma\mathbf{X}[\mathbf{H}^T \circ (1 - \mathbf{H})^T \circ [\mathbf{H}^\dagger(\mathbf{H}\mathbf{X}^T)(\mathbf{X}\mathbf{H}^\dagger) - \mathbf{X}^T(\mathbf{X}\mathbf{H}^\dagger)]
\end{aligned}$$

where γ is a new hyperparameter. That is, in addition to predicting target vectors, we also predict the input vectors themselves in each module of the DSN, which serves to regularize the training.

With all modules of the DSN well initialized and learned, we take the output layer of the top module as the "deep learning features" to feed to the softmax classifier. As reported in the experiments below, for the static pattern classification of MNIST, the gain is relatively minor. But for sequence classification as in the TIMIT phone recognition task, the use of the additional log-linear model is more effective.

Now we turn to the use of K-DSN to generate features for log-linear classifiers. First, we review some practical aspects of tuning the K-DSN parameters. There are many fewer parameters in the K-DSN to tune than the DCN, and there is no need for initialization with often slow, empirical procedures

Table 5.1: Classification error rates using various versions of the DSN in the standard MNIST task (no training data augment with elastic distortion and no convolution)

DSN Models	Error Rate (%)
DSN (one module, no learning of **W**)	1.78
DSN (one module, learning using Eq.(5.16)	1.10
DSN (10 modules, learning using Eq.(5.16))	0.83
DSN (10 modules, learning using Eq.(5.17))	0.79
DSN (10 modules, learning using Eq.(5.17) & adding softmax)	0.77

Table 5.2: Phone error rates using various versions of the DSN in the standard TIMIT phone recognition task

DSN Models	Error Rate (%)
DSN (one module, no learning of **W**)	33.0
DSN (one module, learning using Eq.(5.16)	28.0
DSN (15 modules, learning using Eq.(5.16))	24.6
DSN (15 modules, learning using Eq.(5.17))	23.0
DSN (15 modules, learning using Eq.(5.17) & adding softmax)	20.5

for training the RBM. Importantly, it was found that regularization plays a more important role in K-DCN than in the DSN. The effective regularization schedules that have been developed can often have intuitive insight and can be motivated by optimization tricks. In fact, the regularization procedure shown effective has been motivated by the R-prop algorithm reported in the neural network literature (67).

Further, it was found empirically that in contrast to the DSN, the effectiveness of the K-DSN does not require careful data normalization and it can more easily handle mixed binary and continuous-valued inputs without data and output calibration.

The attractiveness of the K-DSN over the DSN lies partly in its small number of hyperparameters. This makes it easy to adapt the K-DSN from one domain to another. For example, adapting the kernel smoothing parameter in the Gaussian kernel, σ, can be easily carried out with much smaller amounts of adaptation data than adapting other deep models such as DNNs or DSNs.

Given the K-DSN features from the top module, we can connect them directly to a log-linear model as its inputs to perform classification tasks.

One successful example is given in (29) for the sequence classification task of slot filling in spoken language understanding.

5.4 DNN Features for Conditional-Random-Field Sequence Classifiers

As another prominent example of connecting deep learning features to log-linear models, we use DNN features as the input to a very popular log-linear model — the conditional random field (CRF). The combined model, the DNN-CRF, which can be learning in an end-to-end manner, has been highly successful in speech recognition, performing significantly better than the DNN-HMM (28, 51). In the literature reporting this type of model, the DNN-CRF is also described as the DNN-HMM but the DNN parameters are not learned using the common objective function of cross-entropy but rather using the sequence-level criterion of maximum mutual information (MMI). The latter is also called full-sequence-trained DNN-HMM. Equivalence between the DNN-CRF and the sequence-trained DNN-HMM is well understood, which is elaborated in (45, 28).

The motivations for the DNN-CRF are discussed here. In the DNN-HMM, the DNNs are learned to optimize the per-frame cross-entropy between the target HMM state and the DNN predictions. The transition parameters of the HMM and language model scores can be obtained from an HMM-like approach and be trained independently of the DNN weights. However, it has long been known that sequence classification criteria, which are more directly correlated with the overall word or phone error rate, can be very helpful in improving recognition accuracy (2, 37).

In (28), the DNN-CRF was proposed and developed. It was recognized that the use of cross entropy to train DNN for phone sequence recognition does not explicitly take into account the fact that the neighboring frames have smaller distances between the assigned probability distributions over phone class labels. In the proposed approach, the MMI, or equivalently the full-sequence posterior probability $p(l_{1:T}|v_{1:T})$ of the whole sequence of labels, $l_{1:T}$ given the whole visible feature utterance $v_{1:T}$ (or the hidden feature sequence $h_{1:T}$ extracted by the DNN), is used as the objective function for learning all parameters of the DNN-CRF:

$$p(l_{1:T}|v_{1:T}) = p(l_{1:T}|h_{1:T}) = \frac{exp(\sum_{t=1}^{T} \gamma_{ij}\phi_{ij}(l_{t-1}, l_t) + \sum_{t=1}^{T} \sum_{d=1}^{D} \lambda_{l_t,d} h_{td})}{Z(h_{1:T})},$$

$$(5.18)$$

where the transition feature $\phi_{ij}(l_{t-1}, l_t)$ takes on a value of one if $l_{t-1} = i$ and $l_t = j$, and otherwise takes on a value of zero, where γ_{ij} is the parameter associated with this transition feature, h_{td} is the d-th dimension of the hidden unit value at the t-th frame at the final layer of the DNN, and where D is the number of units in the final hidden layer. Note the objective function of Eqn.(5.18) derived from mutual information (40) is the same as the conditional likelihood associated with a specialized linear-chain conditional random field. Here, it is the top most layer of the DNN below the softmax layer, not the raw speech coefficients of MFCC or PLP, that provides "features" to the conditional random field.

To optimize the log conditional probability $p(l_{1:T}^n | v_{1:T}^n)$ of the n-th utterance, we take the gradient over the activation parameters λ_{kd}, transition parameters γ_{ij}, and the lower-layer weights of the DNN, w_{ij}, according to

$$\frac{\partial \log p(l_{1:T}^n | v_{1:T}^n)}{\partial \lambda_{kd}} = \sum_{t=1}^{T} (\delta(l_t^n = k) - p(l_t^n = k | v_{1:T}^n)) h_{td}^n \qquad (5.19)$$

$$\frac{\partial \log p(l_{1:T}^n | v_{1:T}^n)}{\partial \gamma_{ij}} = \sum_{t=1}^{T} [\delta(l_{t-1}^n = i, l_t^n = j) - p(l_{t-1}^n = i, l_t^n = j | v_{1:T}^n)] \qquad (5.20)$$

$$\frac{\partial \log p(l_{1:T}^n | v_{1:T}^n)}{\partial w_{ij}} = \sum_{t=1}^{T} [\lambda_{l_{td}} - \sum_{k=1}^{K} p(l_t^n = k | v_{1:T}^n) \lambda_{kd}] \times h_{td}^n (1 - h_{td}^n) x_{ti}^n \qquad (5.21)$$

Note that the gradient $\dfrac{\partial \log p(l_{1:T}^n | v_{1:T}^n)}{\partial w_{ij}}$ above can be viewed as back-propagating the error $\delta(l_t^n = k) - p(l_t^n = k | v_{1:T}^n)$, vs. $\delta(l_t^n = k) - p(l_t^n = k | v_t^n)$ in the frame-based training algorithm.

In implementing this learning algorithm, which is iterative, the DNN weights are first pre-learned by optimizing the per-frame cross entropy. The transition parameters are then initialized from the combination of the HMM transition matrices and the "phone language" model scores. These transition parameters are further optimized by tuning the transition features while fixing the DNN weights before the joint optimization. Using the joint optimization with careful scheduling, this full-sequence MMI training is shown to outperform the frame-level training by about 5% relative. Subsequent work on full-sequence MMI training for the DNN-CRF has shown greater success in much larger speech recognition tasks, where lattices need to be invoked when n-gram language models with $n > 2$ are used and more sophisticated optimization methods need to exploited.

5.5   Log-Linear Stacking to Combine Multiple Deep Learning Systems

As the final case study on how log-linear models can be usefully connected to deep learning systems, we combine several deep learning systems' outputs as the features and use a log-linear model either to perform classification or to produce new and more powerful features for a more complex structured classification problem.

For the latter, we describe a method, called log-linear stacking here. Without loss of generality, we take the example of three deep learning systems, whose vector-valued outputs are $\mathbf{X} = [\mathbf{x}_1, ..., \mathbf{x}_i, ..., \mathbf{x}_N]$, $\mathbf{Y} = [\mathbf{y}_1, ..., \mathbf{y}_i, ..., \mathbf{y}_N]$, and $\mathbf{Z} = [\mathbf{z}_1, ..., \mathbf{z}_i, ..., \mathbf{z}_N]$, respectively. The log-linear stacking method combines such outputs, which may be posterior probabilities of discrete classes, according to

$$\mathbf{U} \log \mathbf{x}_i + \mathbf{V} \log \mathbf{y}_i + \mathbf{W} \log \mathbf{z}_i + \mathbf{b} \tag{5.22}$$

where matrices $\mathbf{U}$, $\mathbf{V}$, and $\mathbf{W}$, and vector $\mathbf{b}$ are free stacking parameters to be learned so that combined features approach their given target-class values $\mathbf{T} = [\mathbf{t}_1, ..., \mathbf{t}_i, ..., \mathbf{t}_N]$ in a supervised learning setting.

To learn these free parameters, we can adopt least square error as the cost function, subject to L_2 regularization:

$$E = 0.5 \sum_{i=1}^{T} ||\mathbf{U}\tilde{\mathbf{x}}_i + \mathbf{V}\tilde{\mathbf{y}}_i + \mathbf{W}\tilde{\mathbf{z}}_i + \mathbf{b}||^2 + \lambda_1 ||\mathbf{U}||^2 + \lambda_2 ||\mathbf{V}||^2 + \lambda_3 ||\mathbf{W}||^2 \tag{5.23}$$

where

$$\tilde{\mathbf{x}}_i = \log \mathbf{x}_i, \quad \tilde{\mathbf{y}}_i = \log \mathbf{y}_i, \quad and \quad \tilde{\mathbf{z}}_i = \log \mathbf{z}_i,$$

and λ_1, λ_2 and λ_3 are Lagrange multipliers treated as tunable hyperparameters.

The solution to the problem of minimizing (Eq.5.23) is the following analytical operation involving matrix inversion and vector-matrix multiplication:

$$[\mathbf{U}, \mathbf{V}, \mathbf{W}, \mathbf{b}] = [\mathbf{T}\tilde{\mathbf{X}}', \mathbf{T}\tilde{\mathbf{Y}}', \mathbf{T}\tilde{\mathbf{Z}}', \sum_i \mathbf{t}_i] \, \mathbf{M}^{-1}, \tag{5.24}$$

where

$$
\mathbf{M} = \left[\begin{array}{cccc}
\tilde{\mathbf{X}}\tilde{\mathbf{X}}' + \lambda_1\mathbf{I} & \tilde{\mathbf{Y}}\tilde{\mathbf{X}}' & \tilde{\mathbf{Z}}\tilde{\mathbf{X}}' & \sum_{i=1}^{N}\tilde{\mathbf{x}}' \\
\tilde{\mathbf{X}}\tilde{\mathbf{Y}}' & \tilde{\mathbf{Y}}\tilde{\mathbf{Y}}' + \lambda_2\mathbf{I} & \tilde{\mathbf{Z}}\tilde{\mathbf{Y}}' & \sum_{i=1}^{N}\tilde{\mathbf{y}}' \\
\tilde{\mathbf{X}}\tilde{\mathbf{Z}}' & \tilde{\mathbf{Y}}\tilde{\mathbf{Z}}' & \tilde{\mathbf{Z}}\tilde{\mathbf{Z}}' + \lambda_3\mathbf{I} & \sum_{i=1}^{N}\tilde{\mathbf{z}}' \\
\sum_{i=1}^{N}\tilde{\mathbf{x}}' & \sum_{i=1}^{N}\tilde{\mathbf{y}}' & \sum_{i=1}^{N}\tilde{\mathbf{z}}' & N
\end{array} \right] \tag{5.25}
$$

and where $\mathbf{T} = [\mathbf{t}_1, ..., \mathbf{t}_i, ..., \mathbf{t}_N]$ defined earlier.

The log-linear stacking described above has been applied to combine three deep learning systems (DNN, RNN, and CNN) for speech recognition. The experimental paradigm is shown in Figure 5.4, where the log-linearly assembled features are fed to an HMM sequence decoder. Positive experimental results have been reported recently in (20).

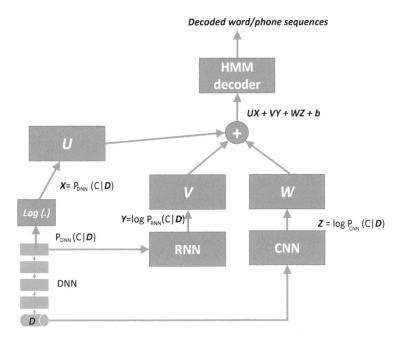

Figure 5.6: Block diagram showing the experimental paradigm for the use of log-linear stacking of three deep learning systems' outputs for speech recognition. The three systems are 1) deep neural network (DNN, left); 2) recurrent neural network which takes the DNN outputs as its inputs (RNN, middle), and 3) Convolutional neural network (CNN, right).

5.6   Discussion and Conclusions

In this chapter, we provide a new way of looking at the DNN and other deep architectures (e.g., the DSN and K-DSN) in terms of their ability to provide high-level features to relatively simple classifiers such as log-linear models, rather than viewing an entire deep learning model as a complex classifier. Since log-linear models can either be a simple static classifier, such as the softmax or maximum entropy models, or be a structured sequence classifier, the deep classifiers based on high-level features provided by the common deep learning systems can be either static or sequential. The parameters of such deep classifiers can be learned either in an end-to-end manner—using backpropagation from the log-linear model's output all the way into the deep learning system, or in two separate stages—keeping the pre-trained deep learning features fixed while learning the log-linear model parameters. The former way of training tends to be more effective in reducing training errors but is also prone to overfitting. The latter training method, on the other hand, would suffer less from overfitting and be most appropriate for multi-task or transfer learning where the in-domain training data are relatively scarce.

We have provided three case studies in this chapter to concretely illustrate the effectiveness of the above paradigm of pre-training deep learning features in constructing deep classifiers grounded on otherwise shallow log-linear modeling. The examples include the use of the DSN and K-DSN to construct a static deep classifier based on the softmax function, the use of the DNN to construct a sequential deep classifier based on the CRF, and the use of aggregated DNN, CNN, and RNN to construct a log-linear stacker. Other case studies, not covered in detail in this chapter but demonstrating similar effectiveness, would include the use of DNN- or DSN-extracted high-level features for RNN-based and CRF-based sequence classifiers.

Returning to the main theme of this book on log-linear modeling, we point out that one major attraction of this modeling paradigm is the flexibility in exploiting a wide range of human-engineered features based on the application-domain knowledge and problem constraints. The paradigm advanced in this chapter is to replace these knowledge-driven features by automatically acquired, data-drive features derived by the DNN while maintaining the same structured sequence discriminative power of log-linear models such as the CRF. However, the obvious downside of this data-driven approach is the lack of freedom to incorporate domain knowledge and problem constraints which are reliable and relevant to the tasks. One promising direction to overcome this weakness is to integrate the DNN feature extractor with deep generative models, which, via the mechanism of probabilistically modeling dependencies among latent variables and the observed data, can naturally embed application-domain constraints in the model space. This provides a more principled way to incorporate domain knowledge than in the feature space as is commonly done in traditional log-linear modeling. Then the integrated DNN feature extractor, which may use appropriately derived activation functions and structured weight matrices to reflect the embedded problem constraints, will more than adequately compensate for the lack of hand-crafted features based on knowledge engineering in the traditional log-linear modeling. This is non-trivial research, attributed largely to intractability in inference algorithms associated with most deep generative models including those for the human speech process (15, 16, 55, 56).

While the generally intractable deep generative models are difficult to integrate with the DNN in principle, simpler and less deep or even shallow generative models may be exploited even though they incorporate domain-specific properties and constraints in a cruder manner. In this case, the shallow models can be stacked layer-by-layer, each "layer" being associated with one iterative step in the easy inference procedure for a shallow generative model, to form a deep architecture. The formation of the deep model may also be accomplished in ways similar to stacking restricted Boltzmann machines to form the deep belief network (43, 44) and to stacking shallow neural nets (with one hidden layer) to form a DSN (30). Then the weights on different layers can be relaxed to be independent of each other and the whole deep network can be trained using the powerful and efficient discriminative algorithm of backpropagation (e.g., (74)). Without running the backpropagation algorithm to full convergence (i.e., early termination) or with the use of appropriate constraints in optimizing the desired objective function (e.g., (9)), domain knowledge and problem constraints established in the generative model before backpropagation will be partially maintained after the discriminative optimization for parameter updates.

In summary, deep learning features, including those provided by the DNN and DSN as exemplified in the case studies presented in this chapter, are powerful features for log-linear modeling in discrimination tasks. They use layer-by-layer structure to automatically extract information from raw data instead of relying on human experts and knowledge workers to design the non-adaptive features based on the same raw data before feeding them into a log-linear model. However, the advantage of automating feature extraction using DNNs carries with it the limitation of not being able to embed reliable domain knowledge often useful or essential for machine learning tasks in hand such as classification. In order to overcome this apparent weakness, a future direction is to seamlessly fuse DNNs and deep generative models, where the former is good at accomplishing the end task's goal via backpropagation-like learning and the latter excels at naturally incorporating application-domain constraints and knowledge in the appropriate model space. After getting the best of both worlds, deep learning systems are expected to advance further from the DNN-centric paradigm described in this chapter.

References

1. T. Anastasakos, J. Mcdonough, R. Schwartz, and J. Makhoul. A compact model for speaker-adaptive training. In *Proc. International Conference on Spoken Language Processing (ICSLP)*, pages 1137–1140, 1996.

2. L. R. Bahl, P. F. Brown, P. V. de Souza, and R. L. Mercer. Maximum mutual information estimation of HMM parameters for speech recognition. In *Proc. International Conference on Acoustics, Speech and Signal Processing (ICASSP)*, 1986.

3. J. Baker, L. Deng, J. Glass, S. Khudanpur, C.-H. Lee, N. Morgan, and D. O'Shgughnessy. Research developments and directions in speech recognition and understanding, part i. *IEEE Signal Processing Magazine*, 26(3):75–80, 2009.

4. J. Baker, L. Deng, J. Glass, S. Khudanpur, C.-H. Lee, N. Morgan, and D. O'Shgughnessy. Updated MINDS report on speech recognition and understanding. *IEEE Signal Processing Magazine*, 26(4):78–85, 2009.

5. Y. Bengio. Learning deep architectures for AI. *Foundations and Trends in Machine Learning*, 2(1):1–127, 2009.

6. A. Biem, S. Katagiri, E. McDermott, and Biing-Hwang Juang. An application of discriminative feature extraction to filter-bank-based speech recognition. *Speech and Audio Processing, IEEE Transactions on*, 9(2): 96–110, Feb 2001.

7. L. Breiman. Stacked regression. *Machine Learning*, 24:49–64, 1996.

8. J. Bridle, L. Deng, J. Picone, H. Richards, J. Ma, T. Kamm, M. Schuster, S. Pike, and R. Reagan. An investigation fo segmental hidden dynamic models of speech coarticulation for automatic speech recognition. *Final Report for 1998 Workshop on Langauge Engineering, CLSP, Johns Hopkins*, 1998.

9. J. Chen and L. Deng. A primal-dual method for training recurrent neural networks constrained by the echo-state property. In *Proc. ICLR*, 2014.

10. R. Chengalvarayan and L. Deng. HMM-based speech recognition using state-dependent, discriminatively derived transforms on mel-warped dft features. *IEEE Transactions on Speech and Audio Processing*, (5): 243256, 1997.

11. R. Chengalvarayan and L. Deng. Use of generalized dynamic feature parameters for speech recognition. *IEEE Transactions on Speech and Audio Processing*, 5(3):232–242, May 1997.

12. William W. Cohen and Vitor Rocha de Carvalho. Stacked sequential learning. In *Proc. IJCAI*, pages 671–676, 2005.

13. G. Dahl, Dong Yu, Li Deng, and Alex Acero. Large vocabulary continuous speech recognition with context-dependent DBN-HMMs. In *Proc. International Conference on Acoustics, Speech and Signal Processing (ICASSP)*, pages 4688–4691, 2011.

14. George E Dahl, Dong Yu, Li Deng, and Alex Acero. Context-dependent pre-trained deep neural networks for large-vocabulary speech recognition. *IEEE Transactions on Audio, Speech and Language Processing*, 20 (1):30–42, 2012.

15. L. Deng. Computational models for speech production. In *Computational Models of Speech Pattern Processing*, pages 199–213. Springer-Verlag, New York, 1999.

16. L. Deng. Switching dynamic system models for speech articulation and acoustics. In *Mathematical Foundations of Speech and Language Processing*, pages 115–134. Springer-Verlag, New York, 2003.

17. L. Deng and J. Chen. Sequence classification using high-level features extracted from deep neural networks. In *Proc. International Conference on Acoustics, Speech and Signal Processing (ICASSP)*, 2014.

18. L. Deng and X. Li. Machine learning paradigms in speech recognition: An overview. *IEEE Transactions on Audio, Speech, and Language Processing,*, 21(5):1060–1089, 2013.

19. L. Deng and D. O'Shaughnessy. *SPEECH PROCESSING — A Dynamic and Optimization-Oriented Approach*. Marcel Dekker Inc, NY, 2003.

20. L. Deng and John Platt. Ensemble deep learning for speech recognition. In *Proc. Annual Conference of International Speech Communication Association (INTERSPEECH)*, 2014.

21. L. Deng and Roberto Togneri. Deep dynamic models for learning hidden representations of speech features. In *Speech and Audio Processing for Coding, Enhancement and Recognition*. Springer, 2014.

22. L. Deng and D. Yu. Deep convex network: A scalable architecture for speech pattern classification. In *Proc. Annual Conference of International Speech Communication Association (INTERSPEECH)*, 2011.

23. L. Deng and D. Yu. *Deep Learning: Methods and Applications.* NOW Publishers, 2014.

24. L. Deng, P. Kenny, M. Lennig, V. Gupta, F. Seitz, and P. Mermelsten. Phonemic hidden markov models with continuous mixture output densities for large vocabulary word recognition. *IEEE Transactions on Acoustics, Speech and Signal Processing*, 39(7):1677–1681, 1991.

25. L. Deng, G. Ramsay, and D. Sun. Production models as a structural basis for automatic speech recognition. *Speech Communication*, 33(2-3): 93–111, Aug 1997.

26. L. Deng, A. Acero, M. Plumpe, and XD. Huang. Large vocabulary speech recognition under adverse acoustic environments. In *Proc. International Conference on Spoken Language Processing (ICSLP)*, pages 806–809, 2000.

27. L. Deng, G. Hinton, and D. Yu. Deep learning for speech recognition and related applications. In *NIPS Workshop*, Whistler, Canada, 2009.

28. L. Deng, M. Seltzer, D. Yu, A. Acero, A. Mohamed, and G. Hinton. Binary coding of speech spectrograms using a deep auto-encoder. In *Proc. Annual Conference of International Speech Communication Association (INTERSPEECH)*, 2010.

29. L. Deng, G. Tur, X. He, and D. Hakkani-Tur. Use of kernel deep convex networks and end-to-end learning for spoken language understanding. In *Proc. IEEE Workshop on Automfatic Speech Recognition and Understanding (ASRU)*, 2012.

30. L. Deng, D. Yu, and J. Platt. Scalable stacking and learning for building deep architectures. In *Proc. International Conference on Acoustics, Speech and Signal Processing (ICASSP)*, 2012.

31. L. Deng, O. Abdel-Hamid, and D. Yu. A deep convolutional neural network using heterogeneous pooling for trading acoustic invariance with phonetic confusion. In *Proc. International Conference on Acoustics, Speech and Signal Processing (ICASSP)*, Vancouver, Canada, May 2013.

32. L. Deng, G. Hinton, and B. Kingsbury. New types of deep neural network learning for speech recognition and related applications: An overview. In *Proc. International Conference on Acoustics, Speech and Signal Processing (ICASSP)*, Vancouver, Canada, May 2013.

33. P. Divenyi, S. Greenberg, and G. Meyer. *Dynamics of Speech Production and Perception.* IOS Press, 2006.

34. F. Flego and M. Gales. Discriminative adaptive training with VTS and JUD. In *Proc. IEEE Workshop on Automfatic Speech Recognition and Understanding (ASRU)*, pages 170–175, 2009.

35. A. Ghoshal, Pawel Swietojanski, and Steve Renals. Multilingual training of deep-neural netowrks. Proc. International Conference on Acoustics, Speech and Signal Processing (ICASSP), 2013.

36. Asela Gunawardana, Milind Mahajan, Alex Acero, and John C Platt. Hidden conditional random fields for phone classification. In *Proc. Annual Conference of International Speech Communication Association (INTERSPEECH)*, pages 1117–1120, 2005.

37. X. He and L. Deng. *DISCRIMINATIVE LEARNING FOR SPEECH RECOGNITION: Theory and Practice.* Morgan and Claypool, 2008.

38. X. He and L. Deng. Speech recognition, machine translation, and speech translation —- A unified discriminative learning paradigm. *IEEE Signal Processing Magazine*, 27:126–133, 2011.

39. X. He and L. Deng. Speech-centric information processing: An optimization-oriented approach. 31, 2013.

40. X. He, L. Deng, and W. Chou. Discriminative learning in sequential pattern recognition — a unifying review for optimization-oriented speech recognition. *IEEE Signal Processing Magazine*, 25(5):14–36, 2008.

41. G. Heigold, H. Ney, P. Lehnen, T. Gass, and R. Schluter. Equivalence of generative and log-linear models. *IEEE Transactions on Audio, Speech and Language Processing*, 19(5):1138 –1148, july 2011.

42. G. Heigold, V. Vanhoucke, A. Senior, P. Nguyen, M. Ranzato, M. Devin, and J. Dean. Multilingual acoustic models using distributed deep neural networks. Proc. International Conference on Acoustics, Speech and Signal Processing (ICASSP), 2013.

43. G. Hinton and R. Salakhutdinov. Reducing the dimensionality of data with neural networks. *Science*, 313(5786):504 – 507, 2006.

44. G. Hinton, S. Osindero, and Y. Teh. A fast learning algorithm for deep belief nets. *Neural Computation*, 18:1527–1554, 2006.

45. G. Hinton, L. Deng, D. Yu, G. Dahl, A. Mohamed, N. Jaitly, A. Senior, V. Vanhoucke, P. Nguyen, T. Sainath, and B. Kingsbury. Deep neural networks for acoustic modeling in speech recognition. *IEEE Signal Processing Magazine*, 29(6):82–97, 2012.

46. J.-T. Huang, Jinyu Li, Dong Yu, Li Deng, and Yifan Gong. Cross-language knowledge transfer using multilingual deep neural network with

shared hidden layers. In *Proc. International Conference on Acoustics, Speech and Signal Processing (ICASSP)*, 2013.

47. B. Hutchinson, L. Deng, and D. Yu. A deep architecture with bilinear modeling of hidden representations: applications to phonetic recognition. In *Proc. International Conference on Acoustics, Speech and Signal Processing (ICASSP)*, 2012.

48. B. Hutchinson, L. Deng, and D. Yu. Tensor deep stacking networks. *IEEE Transactions on Pattern Analysis and Machine Intelligence (PAMI)*, 35(8):1944–1957, 2013.

49. F. Jiao, S. Wang, Chi-Hoon Lee, Russell Greiner, and Dale Schuurmans. Semi-supervised conditional random fields for improved sequence segmentation and labeling. In *Proc. Annual Meeting of the Association for Computational Linguistics (ACL)*, 2006.

50. O. Kalinli, M. L. Seltzer, J. Droppo, and A. Acero. Noise adaptive training for robust automatic speech recognition. *IEEE Transactions on Audio, Speech and Language Processing*, 18(8):1889 –1901, nov. 2010.

51. B. Kingsbury, T. Sainath, and H. Soltau. Scalable minimum bayes risk training of deep neural network acoustic models using distributed hessian-free optimization. In *Proc. Annual Conference of International Speech Communication Association (INTERSPEECH)*, 2012.

52. A. Krizhevsky, I. Sutskever, and G. Hinton. Imagenet classification with deep convolutional neural networks. In *NIPS*, volume 1, page 4, 2012.

53. J. Lafferty, A. McCallum, and F. Pereira. Conditional random fields: Probabilistic models for segmenting and labeling sequence data. In *Proc. International Conference on Machine Learning (ICML)*, pages 282–289, 2001.

54. Q. Le, M. Ranzato, R. Monga, M. Devin, K. Chen, J. Dean G. Corrado, and A. Ng. Building high-level features using large scale unsupervised learning. In *Proc. International Conference on Machine Learning (ICML)*, 2012.

55. L. J. Lee, H. Attias, and L. Deng. Variational inference and learning for segmental switching state space models of hidden speech dynamics. In *Proc. International Conference on Acoustics, Speech and Signal Processing (ICASSP)*, 2003.

56. L. J. Lee, H. Attias, L. Deng, and P. Fieguth. A multimodal variational approach to learning and inference in switching state space models. In *Proc. International Conference on Acoustics, Speech and Signal Processing (ICASSP)*, 2004.

57. H. Lin, L. Deng, D. Yu, Y. Gong, A. Acero, and C.-H. Lee. A study on multilingual acoustic modeling for large vocabulary ASR. In *Proc. International Conference on Acoustics, Speech and Signal Processing (ICASSP)*, pages 4333–4336, 2009.

58. G. Mann and A. McCallum. Generalized expectation criteria for semi-supervised learning of conditional random fields. In *Proc. Annual Meeting of the Association for Computational Linguistics (ACL)*, 2008.

59. A. Mohamed, George E. Dahl, and Geoffrey Hinton. Deep belief networks for phone recognition. In *NIPS Workshop on Deep Learning for Speech Recognition and Related Applications*, 2009.

60. A. Mohamed, D. Yu, and L. Deng. Investigation of full-sequence training of deep belief networks for speech recognition. In *Proc. Annual Conference of International Speech Communication Association (INTERSPEECH)*, 2010.

61. A. Mohamed, G. Dahl, and G. Hinton. Acoustic modeling using deep belief networks. *IEEE Transactions on Audio, Speech and Language Processing*, 20(1):14–22, jan. 2012.

62. A. Mohamed, Geoffrey Hinton, and Gerald Penn. Understanding how deep belief networks perform acoustic modelling. In *Proc. International Conference on Acoustics, Speech and Signal Processing (ICASSP)*, pages 4273–4276, 2012.

63. N. Morgan. Deep and wide: Multiple layers in automatic speech recognition. *IEEE Transactions on Audio, Speech and Language Processing*, 20(1), jan. 2012.

64. J. Picone, S. Pike, R. Regan, T. Kamm, J. Bridle, L. Deng, Z. Ma, H. Richards, and M. Schuster. Initial evaluation of hidden dynamic models on conversational speech. In *Proc. International Conference on Acoustics, Speech and Signal Processing (ICASSP)*, 1999.

65. D. Povey and P. C. Woodland. Minimum phone error and i-smoothing for improved discriminative training. In *Proc. International Conference on Acoustics, Speech and Signal Processing (ICASSP)*, pages 105–108, 2002.

66. L. Rabiner. A tutorial on hidden markov models and selected applications in speech recognition. *Proceedings of the IEEE*, 77(2):257–286, 1989.

67. M. Riedmiller and H. Braun. A direct adaptive method for faster backpropagation learning: The rprop algorithm. In *IEEE Int. Conf. Neural Networks*, pages 586–591, 1993.

68. T. Sainath, Brian Kingsbury, and Bhuvana Ramabhadran. Auto-encoder bottleneck features using deep belief networks. In *Proc. International Conference on Acoustics, Speech and Signal Processing (ICASSP)*, pages 4153–4156, 2012.

69. T. Sainath, Brian Kingsbury, Abdel-rahman Mohamed, and Bhuvana Ramabhadran. Learning filter banks within a deep neural network framework. In *Proc. IEEE Workshop on Automfatic Speech Recognition and Understanding (ASRU)*, 2013.

70. Jürgen Schmidhuber. Deep learning in neural networks: An overview. *CoRR*, abs/1404.7828, 2014.

71. F. Seide, G. Li, and D. Yu. Conversational speech transcription using context-dependent deep neural networks. In *Proc. Annual Conference of International Speech Communication Association (INTERSPEECH)*, August 2011.

72. F. Seide, Gang Li, Xie Chen, and Dong Yu. Feature engineering in context-dependent deep neural networks for conversational speech transcription. In *Proc. IEEE Workshop on Automfatic Speech Recognition and Understanding (ASRU)*, pages 24–29, 2011.

73. K. Stevens. *Acoustic Phonetics*. MIT Press, 2000.

74. V. Stoyanov, A. Ropson, and J. Eisner. Empirical risk minimization of graphical model parameters given approximate inference, decoding, and model structure. *Proc. International Conference on Artificial Intelligence and Statistics (AISTATS)*, 2011.

75. Z. Tuske, M. Sundermeyer, R. Schluter, and H. Ney. Context-dependent MLPs for LVCSR: Tandem, hybrid or both. In *Proc. Annual Conference of International Speech Communication Association (INTERSPEECH)*, 2012.

76. D. H. Wolpert. Stacked generalization. *Neural Networks*, 5(2):241–259, 1992.

77. S. J. Wright, D. Kanevsky, L. Deng, X. He, G. Heigold, and H. Li. Optimization algorithms and applications for speech and language processing. *IEEE Transactions on Audio, Speech, and Language Processing*, 21(11):2231–2243, November 2013.

78. Z. Yan, Qiang Huo, and Jian Xu. A scalable approach to using DNN-derived features in GMM-HMM based acoustic modeling for LVCSR. In *Proc. Annual Conference of International Speech Communication Association (INTERSPEECH)*, 2013.

79. D. Yu and L. Deng. Deep-structured hidden conditional random fields for phonetic recognition. In *Proc. International Conference on Acoustics, Speech and Signal Processing (ICASSP)*, 2010.

80. D. Yu, L. Deng, and Alex Acero. Hidden conditional random field with distribution constraints for phone classification. In *Proc. Annual Conference of International Speech Communication Association (INTERSPEECH)*, pages 676–679, 2009.

81. D. Yu, L. Deng, and George Dahl. Roles of pre-training and fine-tuning in context-dependent DBN-HMMs for real-world speech recognition. In *Proc. Neural Information Processing Systems (NIPS) Workshop on Deep Learning and Unsupervised Feature Learning*, 2010.

82. D. Yu, L. Deng, and F. Seide. The deep tensor neural network with applications to large vocabulary speech recognition. 21(3):388–396, 2013.

83. D. Yu, M. Seltzer, Jinyu Li, Jui-Ting Huang, and Frank Seide. Feature learning in deep neural networks - studies on speech recognition tasks, 2013.

6 Informative Nonstationarity in Paleoclimatological Log-Linear Models

Yuriy S. Polyakov   polyakov@njit.edu
New Jersey Institute of Technology
Newark, NJ, USA

Fabrice Lambert   lambert@dgf.uchile.cl
Center for Climate and Resilience Research, University of Chile
Santiago, Chile

Serge F. Timashev   serget@mail.ru
Karpov Institute of Physical Chemistry, Moscow, Russia
Troitsk, Moscow Region, Russia

We examine the relation between temperature and logarithmized dust flux from the EPICA Dome C ice core (Antarctica) for the last 800,000 years. Our focus is on the nonstationarity analysis of dust flux time series rather than the regression analysis of correlations between temperature and dust flux (conventional method in paleoclimatology). The analysis is performed using flicker-noise spectroscopy, a phenomenological statistical physics framework developed for the analysis of natural signals with stochastically varying components. Our study shows that the logarithmized dust flux time series is statistically nonstationary. It contains alternating intervals of quiescence and high activity. The quiescent intervals appear to be related to glacial conditions in the temperature series. The periods of high activity appear to correspond to warming events, potentially leading to interglacial conditions. These results can be used to explain why significant correlations between temperature and dust flux exist in the glacial conditions and why there is virtually no correlation during interglacial periods.

6.1   Introduction

Paleoclimatology deals with the analysis of climate change on the scale of Earth's entire history. The records for past climate are reconstructed indirectly based on the "proxy" data acquired from paleoclimatic archives, such as ice cores, tree rings, and marine sediment cores. The correlations between the proxy data for modern age and recent climate records are used to build mathematical models that can be applied to infer past climate using the proxy data for the time period of interest. This procedure generally involves a large number of factors with complex interactions and feedback loops.

One of major factors studied in paleoclimatology is the aerosol load of the atmosphere. Mineral dust aerosols are generally emitted from deserts, and their small-sized fractions can travel to distant areas, for example polar regions. This enables paleoclimatologists to examine the dust content in ice cores from polar regions and draw conclusions about atmospheric processes on a global scale. In this study, we examine data from the European Project for Ice Coring in Antarctica (EPICA) Dome C ice core located on the eastern Antarctic plateau. Dust flux in the ice, which is typically measured as the concentration in $(\mathbf{ng\ g}^{-1})$ for soluble dust and particle number concentration in $(\mathbf{P\ ml}^{-1})$ for insoluble dust as a function of ice core depth, is generally sensitive to source emissions and atmospheric cleansing, both of which are linked to the characteristics of the hydrological cycle, that is in turn related to sea surface temperature.

This relationship between dust and temperature is nonlinear: it has multiple modes and thresholds with respect to temperature (7). Due to the low accumulation rate in Antarctica and the log-normal distribution of dust proxy data, the logarithmic values of dust flux are generally used for studying the dust-temperature coupling (2), effectively leading to a log-linear model. Typically, the logarithms of dust fluxes and temperature are uncorrelated for warm periods (interglacial mode) and correlated for relatively cold periods (glacial conditions) (7). Several types of correlations are observed depending on temperature thresholds and temporal proximity to glacial terminations (7, 10). These correlations are established using regression analysis methods.

In this chapter, we look beyond the conventional regression analysis techniques by analyzing the statistical nonstationarity in the high-frequency component of logarithmized dust flux. We show that high nonstationarity on certain time intervals may be related to atmospheric reorganizations that are likely to accompany major climatic transitions. In other words, nonstationarity analysis provides new information on the dynamics of past climate change, which cannot be extracted by regression analysis.

To examine the nonstationarity, we use flicker-noise spectroscopy (FNS), a phenomenological framework for extracting information from time series with stochastically varying components (16, 17, 18). We apply the FNS nonstationarity factor function to identify the time intervals of major rearrangements (within relatively short time intervals) of the complex system under study. The FNS nonstationarity factor was previously used to locate precursors to strong earthquakes (1, 12, 19, 3, 4, 11).

The chapter is structured as follows. Section 6.2 introduces the principles of FNS. Sections 6.3 and 6.4 explain the algorithms for smoothing and evaluating the nonstationarity factor, respectively. Section 6.5 depicts the complete procedure for the nonstationarity analysis of dust flux time series. Section 6.6 briefly describes the experimental data. The results of the analysis and their interpretation are provided in Section 6.7. The main conclusions and directions for future research studies are given in Section 6.8.

6.2 Principles of Flicker-Noise Spectroscopy

Here, we will only deal with the basic FNS relations needed to understand the nonstationarity factor. FNS is described in more detail elsewhere (16, 17, 18, 15, 8).

In FNS, all introduced parameters for signal $V(t)$, where t is time, are related to the autocorrelation function

$$\psi(\tau) = \langle V(t)V(t+\tau)\rangle_{T-\tau}, \tag{6.1}$$

where τ is the time lag parameter ($0 < \tau \leq T_M$) and T_M is the upper bound for τ ($T_M \leq T/2$). This function characterizes the correlation in values of dynamic variable V at higher, $t+\tau$, and lower, t, values of the argument. The angular brackets in relation (6.1) stand for the averaging over time interval $[0, T - \tau]$:

$$\langle(...)\rangle_{T-\tau} = \frac{1}{T-\tau}\int_0^{T-\tau}(...)\,dt. \tag{6.2}$$

The averaging over interval $[0, T - \tau]$ implies that all the characteristics that can be extracted by analyzing functions $\psi(\tau)$ should be regarded as the average values on this interval.

To extract the information contained in $\psi(\tau)$ ($\langle V(t)\rangle = 0$ is assumed), the following transforms, or "projections", of this function are analyzed: cosine transforms ("power spectrum" estimates) $S(f)$, where f is the frequency,

$$S(f) = 2 \int_0^{T_M} \langle V(t)V(t+t_1) \rangle_{T-\tau} \cos(2\pi f t_1) \, dt_1 \tag{6.3}$$

and its difference moments (Kolmogorov transient structure functions) of the second order $\Phi^{(2)}(\tau)$

$$\Phi^{(2)}(\tau) = \left\langle [V(t) - V(t+\tau)]^2 \right\rangle_{T-\tau}. \tag{6.4}$$

Here, we use the quotes for power spectrum because according to the Wiener-Khinchin theorem the cosine (Fourier) transform of autocorrelation function is equal to the power spectral density only for wide-sense stationary signals at infinite integration limits.

The information contents of $S(f)$ and $\Phi^{(2)}(\tau)$ are generally different, and the parameters for both functions are needed to solve parameterization problems. By considering the intermittent character of signals under study, interpolation expressions for the stochastic components $\Phi_s^{(2)}(\tau)$ and $S_s(f)$ of $S(f)$ and $\Phi^{(2)}(\tau)$, respectively, were derived using the theory of generalized functions by (14). It was shown that the stochastic components of structure functions $\Phi^{(2)}(\tau)$ are formed only by jump-like irregularities ("random walks"), and stochastic components of functions $S(f)$, which characterize the "energy side" of the process, are formed by spike-like (inertial) and jump-like irregularities.

6.3 FNS Smoothing Procedure

The analysis of experimental stochastic series often requires the original data to be smoothed. In this study, we apply the "relaxation" procedure proposed for nonstationary signals by (13) based on the analogy with a finite-difference solution of the diffusion equation, which allows one to split the original signal into low-frequency $V_R(t)$ and high-frequency $V_F(t)$ components.

Consider the one-dimensional diffusion equation for V_R :

$$\frac{\partial V_R}{\partial \tau} = \chi \frac{\partial^2 V_R}{\partial t^2} \tag{6.5}$$

with symmetric boundary conditions

$$\frac{\partial V_R}{\partial t} = 0 \quad \text{at} \quad t = 0, \tag{6.6}$$

$$\frac{\partial V_R}{\partial t} = 0 \quad \text{at} \quad t = T \tag{6.7}$$

and initial condition

$$V_R(0) = V(t), \tag{6.8}$$

where χ is a constant diffusion coefficient.

Writing a forward difference for the local term and second-order central difference for the diffusion term, Equation (6.5) gets transformed to

$$\frac{V_k^{i+1} - V_k^i}{\Delta\tau} = \chi \frac{V_{k+1}^i - 2V_k^i + V_{k-1}^i}{(\Delta t)^2}, \tag{6.9}$$

where i is the "time" index and k is the "spatial" index. Here, the subscript R is dropped for simplicity.

After introducing $\omega = \frac{\chi \Delta\tau}{(\Delta t)^2}$, Equation (6.9) can be further transformed to the following explicit finite difference expression:

$$V_k^{i+1} = \omega V_{k+1}^i + (1 - 2\omega) V_k^i + \omega V_{k-1}^i. \tag{6.10}$$

Analogously, the finite difference formulation for the complete problem (6.5)–(6.9) can be written as

$$V_1^{i+1} = (1 - 2\omega) V_1^i + 2\omega V_2^i, \tag{6.11}$$

$$V_k^{i+1} = \omega V_{k+1}^i + (1 - 2\omega) V_k^i + \omega V_{k-1}^i, \tag{6.12}$$

$$V_{N_t}^{i+1} = (1 - 2\omega) V_{N_t}^i + 2\omega V_{N_t-1}^i. \tag{6.13}$$

Here, i is the current iteration number and N_t is the length of the time series. The smoothing procedure (6.11)–(6.13) is unconditionally stable for $\omega < 1/2$, which is the maximum allowed value for ω in the smoothing algorithm. There are two input parameters: the number of iterations i_{max} (largest value for i), which is equivalent to a certain "cutoff" frequency for the time series under study, and the value for ω. Expressions (6.11)–(6.13) are evaluated at each iteration.

In summary, the smoothing procedure finds new values of the signal at every "relaxation" step using its values for the previous time step, yielding the low-frequency component V_R at the end. The high-frequency component V_F is obtained by subtracting V_R from the original signal. Conceptually speaking, this algorithm progressively reduces the local gradients of the "concentration" variable, causing the points in every triplet to come closer to each other.

6.4 FNS Nonstationarity Factor

To analyze the effects of nonstationarity in real processes, we study the dynamics of changes in $\Phi^{(2)}(\tau)$ for consecutive "window" intervals $[t_k, t_k+T]$, where $k = 0, 1, 2, 3, \ldots$ and $t_k = k\Delta T$, that are shifted within the total time interval T_{tot} of experimental time series $(t_k + T < T_{tot})$. The averaging interval T and difference ΔT are chosen based on the physical understanding of the problem in view of the suggested characteristic time of the process, which is the key parameter of system evolution.

The FNS nonstationarity factor $C_J(t_k)$ is defined as

$$C_J(t_k) = 2 \cdot \frac{Q_k^J - P_k^J}{Q_k^J + P_k^J} \cdot \frac{T}{\Delta T}, \tag{6.14}$$

$$Q_k^J = \frac{1}{\alpha T^2} \int_0^{\alpha T} \int_{t_k}^{t_k+T} [V_J(t) - V_J(t+\tau)]^2 \, dt \, d\tau, \tag{6.15}$$

$$P_k^J = \frac{1}{\alpha T^2} \int_0^{\alpha T} \int_{t_k}^{t_k+T-\Delta T} [V_J(t) - V_J(t+\tau)]^2 \, dt \, d\tau. \tag{6.16}$$

Here, J indicates which function $V_J(t)$ $(J = R, F$ or $G)$ is used, and the subscripts R, F, and G refer to the low-frequency (regular) component, high-frequency (fluctuation) component, and unfiltered signal, respectively. Note that the integrands in Eqs. (6.15)–(6.16) correspond to the structural function $\Phi_J^{(2)}(\tau)$ given by Eq. (6.4).

The FNS nonstationarity factor in discrete for is written as $(b = \lfloor \Delta T / \Delta t \rfloor$, $N_1 = \lfloor \alpha N \rfloor)$:

$$C_J(t_k) = 2 \cdot \frac{Q_k^J - P_k^J}{Q_k^J + P_k^J} \Big/ \frac{\Delta T}{T}, \tag{6.17}$$

$$Q_k^J = \frac{1}{N_1} \sum_{n_\tau=1}^{N_1} \frac{1}{N - n_\tau} \sum_{m=1+kb}^{N-n_\tau+kb} [V_J(m) - V_J(m+n_\tau)]^p, \tag{6.18}$$

$$P_k^J = \frac{1}{N_1} \sum_{n_\tau=1}^{N_1} \frac{1}{N - n_\tau} \sum_{m=1+kb}^{N-n_\tau+(k-1)b} [V_J(m) - V_J(m+n_\tau)]^p. \tag{6.19}$$

Note that functions $\Phi_j^{(2)}(\tau)$ can be reliably evaluated only on the τ interval of $[0,\, \alpha T]$ that is less than half the averaging interval T; i.e., $\alpha < 0.5$.

In earthquake prediction studies, the phenomenon of "precursor" occurrence is assumed to be related to abrupt changes in functions $\Phi^{(2)}(\tau)$ when the upper bound of the interval $[t_k, t_k + T]$ approaches the time moment t_c of a catastrophic event accompanied by total system reconfiguration on all space scales. Graphically, this corresponds to peaks in the plots of nonstationarity factor.

6.5 Experimental Data

The EPICA Dome C ice core was drilled in East Antarctica (75°06′ S; 123°21′ E) and covers the last 800,000 years (5). From a depth of 24.2m down to 3200m, a Continuous Flow Analysis (CFA) system (9) was applied to measure, among others, Ca^{2+}, Na^+, and dust particles. The data gathered with this method have a nominal depth resolution of ∼1 cm, taking dispersion in the CFA system into account, which corresponds to a formal sub-annual temporal resolution at the top and up to ∼25 years at the bottom of the ice core. Practically, surface snow mixing and dispersion in the ice result in a lower effective temporal resolution.

In this study, we examine the dust flux time series derived using the principal component analysis of three complete datasets recorded at the EPICA Dome C ice core: soluble Ca^{2+} and non-sea-salt Ca^{2+} concentrations, as well as insoluble dust particle numbers (6). This dataset, hereafter referred to as PC1, was calibrated to dust mass flux units using a two-sided regression analysis between PC1 and dust flux measurements from standard Coulter Counter analysis. More detailed description is provided elsewhere (6).

6.6 Procedure for Analysis

The PC1 time series contains several small gaps (generally not exceeding few sampling points). These missing values were estimated using a linear interpolation between the values adjacent to the gap. Then the time series was ordered chronologically (instead of the standard order by age used in paleoclimatology). The PC1 dust flux values were next logarithmized (due to the low accumulation rate in Antarctica and the log-normal distribution of dust proxy data) to study its correlation with temperature. As a first step, we applied a moving average operation to the time series using 5-point subsets to minimize the effect of single-point spikes. The low-frequency part of the PC1 dust flux is dominated by orbital frequencies (due to eccentricity, obliquity, and precession parameters of Earth's orbit around the Sun) and not relevant to high-frequency changes in Earth's atmosphere. It was therefore excluded using the FNS smoothing procedure (see chapter 6.3), in which we chose the number of iterations i_{max} in such a way that it corresponds to an effective "cuttoff" frequency of $(5 \text{ kyr})^{-1}$. Finally, the nonstationarity factors C_F for different values of averaging interval T were computed, and the resulting time series were reordered by age.

To estimate the significance thresholds corresponding to background noise, the same procedure that was applied to the logarithmized dataset of PC1 was performed for synthetic Gaussian noise (in view of the fact that the power spectrum estimate for the high-frequency component of logarithmized dust flux has a slope relatively close to 0). The histogram of C_F for Gaussian noise was then fitted to the Burr distribution (FNS nonstationarity factor generally has a skewed asymetric distribution that can be well-approximated by the Burr distribution). The 95% and 97.5% significance thresholds were computed for the Gaussian noise signal using the cumulative distribution function. The ratio of the count of values above the threshold for PC1 to the count for Gaussian noise was calculated to check the statistical significance.

6.7   Results and Discussion

Figures 6.1 and 6.2 illustrate the time series for temperature (5), logarithmized PC1, and its low-frequency and high-frequency components for the last 800 kyr and 100 kyr, respectively. It can be seen that temperature and logarithmized PC1 (as well as its low-frequency part) are generally anticorrelated for glacial (low temperature) intervals. During interglacial intervals (highest temperatures), temperature and logarithmized PC1 are much less correlated. This relationship can be examined by regression methods and is discussed in detail elsewhere (7).

Our focus is on the behavior of the high-frequency component, which in this case includes the frequencies between $\sim (5,000 \text{ yr})^{-1}$ and $\sim (50 \text{ yr})^{-1}$. The lower frequency bound corresponds to the "cutoff" for the FNS smoothing procedure, and the upper frequency bound is obtained as a result of applying the moving average procedure on 5-point subsets. Although it is hard to see any direct correlations between temperature and the high-frequency component of logarithmized PC1, one can observe some apparent nonstationarity in the high-frequency component displayed in Fig. 6.1. This nonstationarity can be examined using the FNS nonstationarity factor.

Figures 6.3 and 6.4 show temperature along with the nonstationarity fac-

Figure 6.1: Temperature, logarithmized dust flux at EPICA Dome C, and its low-frequency and high-frequency components for the last 800 kyr.

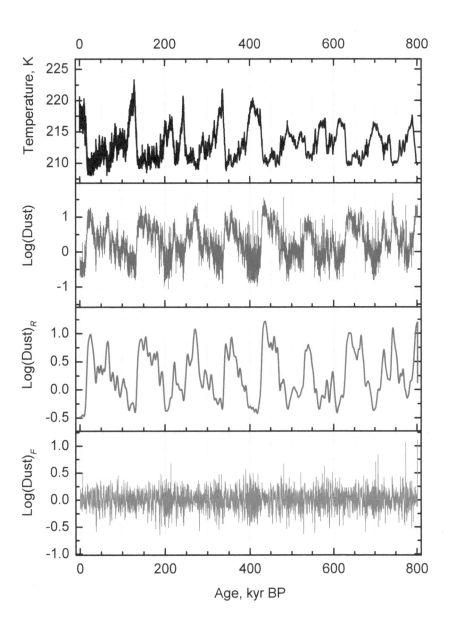

Figure 6.2: Temperature, logarithmized dust flux at EPICA Dome C, and its low-frequency and high-frequency components for the last 100 kyr.

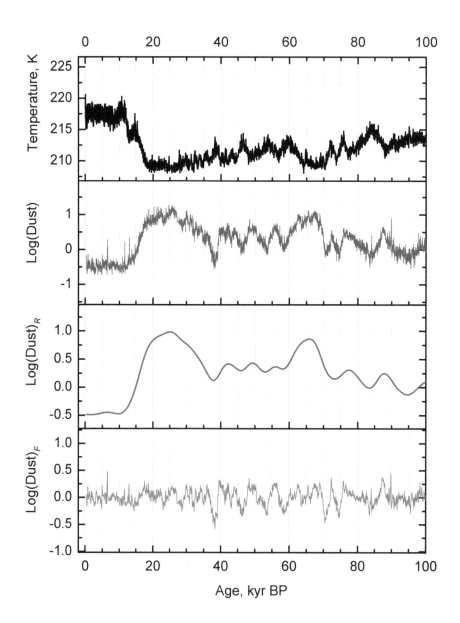

tor for the high-frequency component of logarithmized PC1 at the averaging intervals of 12.5, 25, and 37.5 kyr for the last 800 kyr and 100 kyr, respectively. Each plot for nonstationarity factor C_F also displays two thresholds, 95% (lower) and 97.5% (upper), estimated using the analysis of the nonstationarity factor for Gaussian noise in terms of the Burr distribution. The value of the 95% threshold for C_F is 3.80–3.84 for all three values of the averaging intervals. The values for the 97.5% threshold lie in the range from 5.70 to 5.76. The ratios of the count of points above the 95% threshold for PC1 to the count for Gaussian noise are 1.67:1, 1.40:1, and 1.29:1 for T=12.5, 25, and 37.5 kyr, respectively. In the case of the 97.5% threshold, the ratios are 2.91:1, 2.52:1, and 2.12:1, respectively. The highest values of the ratio correspond to the smallest averaging interval (12.5 kyr).

Figure 6.3 shows that higher values of the averaging interval T lead to more non-uniform distribution of nonstationarity factor. There are intervals of high activity along with quiescent intervals. The plot for $T = 37.5$ kyr demonstrates that quiescent intervals correspond to glacial conditions while the intervals of high activity appear to happen when substantial warming is observed. Physically, this may imply that glacial climate is generally stable until some perturbation (warming) occurs. The perturbation may grow into a large-scale instability, possibly associated with an atmospheric reorganization at multiple scales. This allows us to explain why when the climate system is stable, clear anticorrelations between temperature and logarithmized dust are observed; and then when the climate system becomes unstable, the correlations start to disappear (7).

Figure 6.4 illustrates the intermittence between quiescent and active periods in the plots of nonstationarity factor C_F for the last 100 kyr. For instance, there is a quiescent interval in the C_F plots between 20 kyr and 38 kyr before present (BP) for all values of T, which coincides with full glacial conditions (temperature values do not exceed 212 K). However, there are elevated levels of nonstationarity in the C_F plot between 60 and 38 kyr BP, which corresponds to a period with large temperature variations. Especially during the last glacial-interglacial transition (10-20 kyr BP), it can be seen that C_F at T=12.5 kyr is most informative. In this case it is due to the fact that the time scale of individual features (climatic events) in the plot of temperature is much smaller for the interval from 0 to 100 kyr BP (Fig. 6.4) as compared to the range from 0 to 800 kyr BP (Fig. 6.3). If one looks at a more granular level, the values of the averaging interval should be further reduced to a level comparable to the characteristic time-scale of observed climatic events.

Figure 6.3: Temperature and nonstationarity factor for the high-frequency component of logarithmized PC1 at the averaging intervals of 12.5, 25, and 37.5 kyr for the last 800 kyr. Horizontal lines in each plot for C_F correspond to 95% (lower) and 97.5% (upper) thresholds for Gaussian noise.

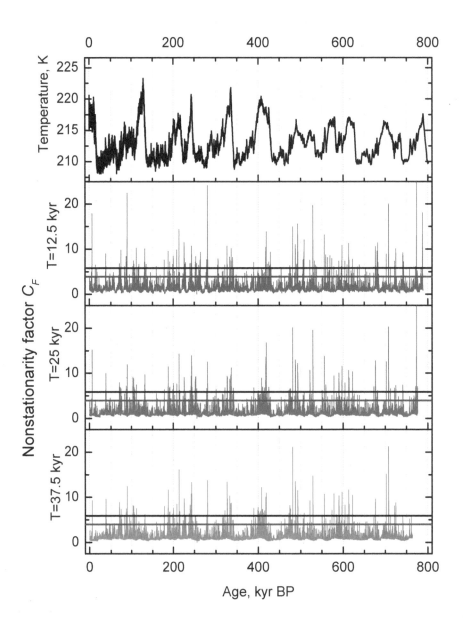

Figure 6.4: Temperature and nonstationarity factor for the high-frequency component of logarithmized PC1 at the averaging intervals of 12.5, 25, and 37.5 kyr for the last 100 kyr. Horizontal lines in each plot for C_F correspond to 95% (lower) and 97.5% (upper) thresholds for Gaussian noise.

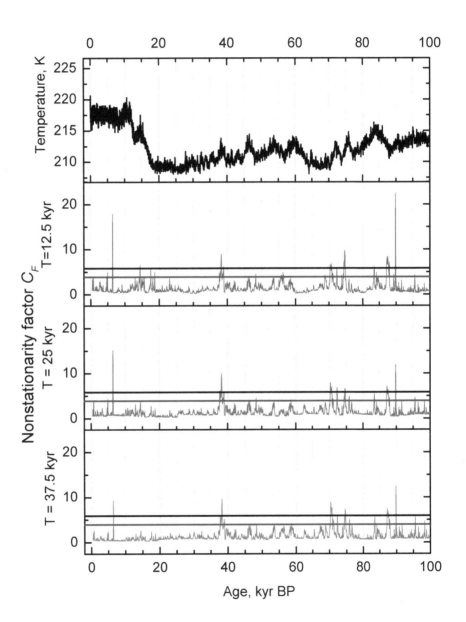

6.8 Concluding Remarks

Our study shows that the dust flux time series derived from the EPICA Dome C ice core data is statistically nonstationary. It contains alternating intervals of quiescence and high activity. The quiescent intervals appear to be related to cold conditions in the temperature series (5), whereas periods of high activity seem to correspond to warming events, potentially leading to interglacial conditions. It is generally thought there is a significant correlation between dust flux and temperature records during glacial periods and virtually no correlation during interglacial periods (7). Our results suggest that the climate system is stable during cold times (quiescent intervals), which is why the correlation is observed. Conversely, the climatic system seems to be perturbed by warming and lose its stability (high activity intervals), thus breaking down the causal link between Southern Hemisphere surface temperature and dust flux in Antarctica.

The main goal of this study is to demonstrate that nonstationarity analysis and particularly the FNS nonstationarity factor may provide new information in the analysis of loglinear models used in paleoclimatology. This information cannot be acquired using regression methods but may explain why certain correlations are present or absent. The information from nonstationarity analysis may also help in developing climate models that study the onset of unstable climatic modes. To form a more complete picture of how nonstationary processes in atmospheric signals interact with temperature, it is necessary to study other climate proxies for Antarctica as well as other geographic regions. One would also need to analyze the time series at different scales (using different values of the averaging interval in evaluating the FNS nonstationarity factor) and examine the behavior for various glacial-interglacial climatic cycles, as well as millennial-scale oscillations in both hemispheres.

References

1. A. V. Descherevsky, A. A. Lukk, A. Y. Sidorin, G. V. Vstovsky, and S. F. Timashev. Flicker-noise spectroscopy in earthquake prediction research. *Nat. Hazards Earth Syst. Sci.*, 3:159–164, 2003.

2. H. Fischer, M.-L. Siggaard-Andersen, U. Ruth, R. Rthlisberger, and E. Wolff. Glacial/interglacial changes in mineral dust and sea-salt records in polar ice cores: Sources, transport, and deposition. *Reviews of Geophysics*, 45(1):RG1002, 2007.

3. M. Hayakawa and S. F. Timashev. An attempt to find precursors in the ulf geomagnetic data by means of flicker noise spectroscopy. *Nonlinear Proc. Geophys.*, 13(3):255–263, 2006.

4. Y. Ida, M. Hayakawa, and S. Timashev. Application of different signal analysis methods to the ulf data for the 1993 guam earthquake. *Nat. Hazards Earth Syst. Sci.*, 7(4):479–484, 2007.

5. J. Jouzel, V. Masson-Delmotte, O. Cattani, G. Dreyfus, S. Falourd, G. Hoffmann, B. Minster, J. Nouet, J. M. Barnola, J. Chappellaz, H. Fischer, J. C. Gallet, S. Johnsen, M. Leuenberger, L. Loulergue, D. Luethi, H. Oerter, F. Parrenin, G. Raisbeck, D. Raynaud, A. Schilt, J. Schwander, E. Selmo, R. Souchez, R. Spahni, B. Stauffer, J. P. Steffensen, B. Stenni, T. F. Stocker, J. L. Tison, M. Werner, and E. W. Wolff. Orbital and millennial antarctic climate variability over the past 800,000 years. *Science*, 317(5839):793–796, 2007.

6. F. Lambert, M. Bigler, J. P. Steffensen, M. Hutterli, and H. Fischer. Centennial mineral dust variability in high-resolution ice core data from dome c, antarctica. *Climate of the Past*, 8(2):609–623, 2012.

7. F. Lambert, B. Delmonte, J. R. Petit, M. Bigler, P. R. Kaufmann, M. A. Hutterli, T. F. Stocker, U. Ruth, J. P. Steffensen, and V. Maggi. Dust-climate couplings over the past 800,000 years from the epica dome c ice core. *Nature*, 452:616–619, 2008.

8. Y. S. Polyakov, J. Neilsen, and S. F. Timashev. Stochastic variability in x-ray emission from the black hole binary GRS 1915+105. *Astron. J.*, 143(6):148, 2012.

9. R. Röthlisberger, M. Bigler, M. Hutterli, S. Sommer, B. Stauffer, H. G. Junghans, and D. Wagenbach. Technique for continuous high-resolution analysis of trace substances in firn and ice cores. *Environmental Science and Technology*, 34(2):338–342, 2000.

10. R. Röthlisberger, M. Mudelsee, M. Bigler, M. de Angelis, H. Fischer, M. Hansson, F. Lambert, V. Masson-Delmotte, L. Sime, R. Udisti, and E. W. Wolff. The southern hemisphere at glacial terminations: insights from the dome c ice core. *Climate of the Past Discussions*, 4(3):761–789, 2008.

11. G. V. Ryabinin, Y. S. Polyakov, V. A. Gavrilov, and S. F. Timashev. Identification of earthquake precursors in the hydrogeochemical and geoacoustic data for the kamchatka peninsula by flicker-noise spectroscopy. *Natural Hazards and Earth System Science*, 11(2):541–548, 2011.

12. L. Telesca, V. Lapenna, S. Timashev, G. Vstovsky, and G. Martinelli.

Flicker-noise spectroscopy: a new approach to investigate the time dynamics of geoelectrical signals measured in seismic areas. *Phys. Chem. Earth.*, 29:389–395, 2004.

13. S. Timashev and G. Vstovskii. Flicker-noise spectroscopy for analyzing chaotic time series of dynamic variables: Problem of signal-to-noise relation. *Russ. J. Electrochem.*, 39(2):141–153, 2003.

14. S. F. Timashev. Flicker noise spectroscopy and its application: Information hidden in chaotic signals. *Russ. J. Electrochem.*, 42:424, 2006.

15. S. F. Timashev, O. Y. Panischev, Y. S. Polyakov, S. A. Demin, and A. Y. Kaplan. Analysis of cross-correlations in electroencephalogram signals as an approach to proactive diagnosis of schizophrenia. *Physica A*, 391(4):1179, 2012.

16. S. F. Timashev and Y. S. Polyakov. Review of flicker noise spectroscopy in electrochemistry. *Fluct. Noise Lett.*, 7:R15–R47, 2007.

17. S. F. Timashev and Y. S. Polyakov. Analysis of discrete signals with stochastic components using flicker noise spectroscopy. *Int. J. Bifurcation Chaos*, 18:2793–2797, 2008.

18. S. F. Timashev, Y. S. Polyakov, P. I. Misurkin, and S. G. Lakeev. Anomalous diffusion as a stochastic component in the dynamics of complex processes. *Phys. Rev. E*, 81:041128-1–041128-17, 2010.

19. G. V. Vstovsky, A. V. Descherevsky, A. A. Lukk, A. Y. Sidorin, and S. F. Timashev. Search for electric earthquake precursors by the method of flicker-noise spectroscopy. *Izv.-Phys. Soild Earth*, 41:513–524, 2005.

7   Log-Linear Quandloids and Information Fusion

Daniel Moskovich   dmoskovich@gmail.com
School of Physical and Mathematical Sciences, Nanyang Technological University
Singapore

Avishy Carmi   avcarmi@bgu.ac.il
Department of Mechanical Engineering, Ben-Gurion University
Israel

We provide an axiomatic characterization of information fusion, on the basis of which we define an information fusion tangle. *Our construction is reminiscent of tangle diagrams in low dimensional topology. Information fusion tangles come equipped with a natural notion of* equivalence. *Equivalent information fusion tangles 'contain the same information' but differ by finite sequences of moves taken from a small set of local moves, each of which replaces one subtangle by another subtangle.*

7.1   Introduction

Information fusion is an umbrella term for concepts and methodologies whose primary goal is to integrate pieces of information from diverse sources. An information fusion algorithm is a vital ingredient of virtually any autonomous system with sensing capabilities (3, 5, 7, 8). Some other applications are sensor networks (9, 15), biometrics (13), and intelligent decision support systems (1, 16). Here we consider pieces of information whose correlations are unknown and some of which may originate from unreliable sources. 'Pieces of information' may refer to estimates of unknown parameters or state variables, or to other related statistical measures such as (unnormalized) probability density functions, Fisher information and Shannon entropy.

We algebraically axiomatize information fusion (Definition 7.1), and we define an *information fusion tangle* (Definition 7.6). Information fusion tangles admit a natural notion of equivalence (7.5). When one or more pieces of information are faulty (*e.g.* biased or inconsistent), the faults will propagate differently in equivalent tangles. Depending on where the faults lie, an information fusion tangle may be configured to its 'least faulty' equivalent configuration. Examples are given in Sections 7.2 and 7.6.

Our notion of equivalence parallels the notion of *ambient isotopy* in knot theory. As such, it represents a link between information fusion and low dimensional topology. We do not discuss it in this note, but low dimensional topology provides insight into the characteristic quantities or *invariants* of information fusion tangles, such as how much information is required to uniquely specify the tangle (its *capacity*) and how many 'independent subtangles' the tangle contains (its *complexity*).

7.2 An Illustrative Example

7.2.1 The example

This example illustrates two different *information fusion tangles* for fusing estimates. These tangles are *equivalent* in the sense that any information in one tangle can uniquely be recovered from the information in the other, but they differ in the consistency of intermediate fused estimates. Thus, one tangle might be 'better' than the other.

Consider three estimators $\hat{X}_0$, $\hat{X}_1$ and $\hat{X}_2$ for the same random variable X. The correlation between these estimators is unknown. We are also provided with estimators C_0, C_1, and C_2 for the error covariances:

$$\operatorname{cov}\left[X - \hat{X}_i\right] \overset{\text{def}}{=} E\left[(X - \hat{X}_i)(X - \hat{X}_i)^T\right] - E\left[X - \hat{X}_i\right]E\left[X - \hat{X}_i\right]^T$$
$$i = 0, 1, 2, \qquad (7.1)$$

where $E\{\cdot\}$ denotes the expectation taken with respect to the underlying joint probability distribution.

A *consistent estimator* $(\hat{X}_i,\ C_i)$, also called a *conservative estimator*, is one which is not *too* optimistic about its belief of what the value of X is (4):

$$C_i \geq \operatorname{cov}\left\{X - \hat{X}_i\right\} \qquad (7.2)$$

i.e. the matrix difference $C_i - \text{cov}\left\{X - \hat{X}_i\right\}$ is positive semi-definite, or equivalently the *precision ellipse* of the $(\hat{X}_i, \; C_i)$ contains inside it the precision ellipse of X. Conversely, an inconsistent estimator is one with lower variance('more optimistic') than the true variance 'in some direction'. We would like all estimators to be consistent because inconsistent estimators may diverge and cause errors.

Covariance intersection (CI) provides a method for fusing **a pair** of consistent estimates whose correlations are unspecified (4, 14). The working principle of CI is that if $\hat{X}$ and $\hat{X}'$ are consistent estimators of X, then so is their convex combination (the proof is provided in (4)):

$$\hat{X}_a = (1 - s)C_aC_0^{-1}\hat{X}_0 + sC_aC_1^{-1}\hat{X}_1, \tag{7.3a}$$

$$C_a^{-1} = (1 - s)C_0^{-1} + sC_1^{-1}, \tag{7.3b}$$

where $s \in [0, 1)$ is a weight parameter. Different choices of the parameter s can be adopted to optimize the update with respect to different performance criteria.

Our goal is to fuse **the triple** $(\hat{X}_0, C_0)$, $(\hat{X}_1, C_1)$, and $(\hat{X}_2, C_2)$ to obtain a single consistent estimator for X. Two fusion schemes are:

1. First fuse $(\hat{X}_0, C_0)$ with $(\hat{X}_1, C_1)$ using a parameter s, then fuse the resulting estimator with $(\hat{X}_2, C_2)$ using a parameter t.

2. First fuse both $(\hat{X}_0, C_0)$ and $(\hat{X}_1, C_1)$ with $(\hat{X}_2, C_2)$ using the parameter t. Then fuse the resulting fused estimators using the parameter s.

Later, we will represent these two fusion schemes by Figure 7.3. Pairs are fused using CI using the same weights s and t. A short computation confirms that both of the above fusion schemes for a consistent triple $(\hat{X}_0, C_0)$, $(\hat{X}_1, C_1)$, and $(\hat{X}_2, C_2)$ result in the same consistent estimator for X.

But what if $(\hat{X}_0, C_0)$ and $(\hat{X}_2, C_2)$ were consistent, but $(\hat{X}_1, C_1)$ was inconsistent? Then both fusion schemes **ultimately** lead to the same estimate of X, which is consistent for an appropriate choice of s and of t. But **at an intermediate stage**, the first fusion scheme involves the inconsistent estimate "$(\hat{X}_0, C_0)$ fused with $(\hat{X}_1, C_1)$ with parameter s", whereas all intermediate fused estimates in the second scheme are consistent. In this case the second fusion scheme is *better* than the first. Conversely, if $(\hat{X}_1, C_1)$ is highly consistent compared to $(\hat{X}_0, C_0)$ and $(\hat{X}_2, C_2)$, then the first fusion scheme would be better than the second.

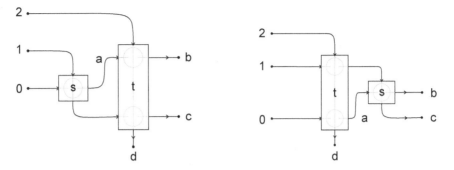

Figure 7.1: First scheme.   **Figure 7.2**: Second scheme.

Figure 7.3: Subfigures 7.1 and 7.2 represent different schemes to fuse estimates $(\hat{X}_0, C_0)$, $(\hat{X}_1, C_1)$, and $(\hat{X}_2, C_2)$ to a consistent estimate for X.

7.2.2   The information fusion tangles of our example

This section constitutes a preliminary discussion of the concept of an information fusion tangle in the context of our example, in anticipation of its definition in Section 7.4. An *information fusion tangle* is made up of labeled directed edges whose incident nodes may either be fixed points in the plane called *endpoints* or boxes called *interactions*. Each interaction involves one distinguished edge called the *agent* which goes 'all the way though' vertically and 'emerges on the other side' (*i.e.* we identify the edge incident to the top to the bottom of the interaction). Other edges are called *patients*, and come in pairs— there is one patient coming into the interaction for each patient going out of the interaction. An agent is an abstraction of an estimator whose value we retain after estimator fusion. One patient abstracts another estimator, and the other abstracts the result of information fusion.

Figure 7.3 contains two subfigures, representing the two information fusion schemes in Section 7.2.1. Each have three *input edges*, labeled 0, 1, 2, denoting estimates $(\hat{X}_0, C_0)$, $(\hat{X}_1, C_1)$, and $(\hat{X}_2, C_2)$ correspondingly, and a distinguished output edge labeled b in 7.1 and $\bar{b}$ in 7.2. Their interactions are labeled by operations $\triangleright_s$ for some $s \in \mathbb{R}$, which represents an application of the fusion rule (7.3) with weight s. Explicitly, agents are weighted s and patients are weighted $1 - s$.

A *fusion process* is a sequence of estimates stored in edges going from an initial to a terminal edge. Subfigure 7.1 contains three fusion processes, labeled $\{0, a, b\}$, $\{1, c\}$, and $\{2\}$. We write $0 \triangleright_s 1$ for the fusion of the estimate $(\hat{X}_0, C_0)$ with $(\hat{X}_1, C_1)$ according to the fusion rule (7.3) with a weighting parameter s. The resulting estimate $(\hat{X}_a, C_a)$ is then passed on to the second interaction where it fuses with $(\hat{X}_2, C_2)$ with parameter t, and resulting estimate $(0 \triangleright_s 1) \triangleright_t 2$ is stored in b.

The fusion process $\{0, a, b\}$ reads:

$$0 \to \underbrace{0 \rhd_s 1}_{a} \to \underbrace{(0 \rhd_s 1) \rhd_t 2}_{b} \tag{7.4}$$

This expression corresponds to the former information fusion scheme of Section 7.2.2.

Direct computation gives:

$$\hat{X}_b = (1-t)(1-s)C_bC_0^{-1}\hat{X}_0 + s(1-t)C_bC_1^{-1}\hat{X}_1 + tC_bC_2^{-1}\hat{X}_2, \tag{7.5a}$$

$$C_b^{-1} = (1-t)(1-s)C_0^{-1} + s(1-t)C_1^{-1} + tC_2^{-1} \tag{7.5b}$$

The tangle of Subfigure 7.2 differs only in its first process, which reads:

$$0 \to \underbrace{0 \rhd_t 2}_{\bar{a}} \to \underbrace{(0 \rhd_t 2) \rhd_s (1 \rhd_t 2)}_{\bar{b}} \tag{7.6}$$

This expression corresponds to the latter information fusion scheme of Section 7.2.2.

By direct computation, we verify indeed that:

$$\bar{b} = (0 \rhd_t 2) \rhd_s (1 \rhd_t 2) = (0 \rhd_s 1) \rhd_t 2 = b \tag{7.7}$$

so that $\hat{X}_{\bar{b}} = \hat{X}_b$ and $C_{\bar{b}} = C_b$.

Thus, the two subfigures of Figure 7.3 share the same parameters s, t and the same initial and terminal colours. They differ only at an intermediate edge, whose label is $a = 0 \rhd_s 1$ in one subfigure and is $\bar{a} = 0 \rhd_t 2$ in the other.

Above we described a pair of equivalent information fusion tangles which have different local performance. We may choose the 'better' of these two tangles relative to known (in)consistencies of the initial estimators. In future sections we consider the general case.

7.3 Algebra of Information Fusion

Consider a sensor network (9, 15). In distributed information fusion architectures, nodes behave as intelligent proxies fusing raw measurements streaming from their sensors with information received from neighboring nodes (15). Fusion may be carried out within a node using a statistical filtering algorithm, *e.g.* the Kalman filter. Such algorithms normally use cross dependencies between the incoming pieces of information (*i.e.* raw measurements and estimates from other nodes). However, large scale and complex networks generally inhibit calculations of the required statistical interdependencies among nodes (4). Covariance intersection (CI) does not require knowledge of correlations, hence it is suitable for fusion in large scale settings (4, 14). Definition 7.1 abstracts the key properties of CI to axiomatize information fusion. Chief among these is self-distributivity or *no double counting* which guarantees equality of outputs of the two tangles of Section 7.2.

7.3.1 Algebraic axiomatization

In this section we formulate an algebra structure of 'information' subject to a binary 'update' operation. We name such a structure a *quandloid*, a portmanteau of "quandle" and "groupoid".

Definition 7.1 (Quandloid). *A quandloid is a set Q, whose elements, called colours, represent pieces of information, together with a set B of partially-defined binary operations representing 'updates', satisfying the following properties:*

Coherence of information *A piece of information may update itself, and this neither generates new information nor loses information. Symbolically:*

$$a \triangleright a = a \qquad \forall a \in Q, \quad \forall \triangleright \in B. \tag{7.8}$$

Causal invertibility *All update operations are left invertible. Thus, any input can uniquely be recovered from its corresponding output together with the agent. Symbolically, for every $\triangleright \in B$ there exists an 'left-inverse operation' $\triangleleft \in B$ such that:*

$$(a \triangleright b) \triangleleft b = a \qquad \forall a, b \in Q, \quad \forall \triangleright \in B. \tag{7.9}$$

More precisely, if $a \triangleright b$ exists then $(a \triangleright b) \triangleleft b$ exists and equals a.

No double counting *Updating $a \triangleright_2 c$ by $b \triangleright_2 c$ with $\triangleright_1$ gives the same result as the updated piece of information $a \triangleright_1 b$ by c with $\triangleright_2$. Thus c counts towards the final result only once, and there is no redundancy. Symbolically:*

$$(a \triangleright_1 b) \triangleright_2 c = (a \triangleright_2 c) \triangleright_1 (b \triangleright_2 c) \qquad \forall a, b, c \in Q, \quad \forall \triangleright_1, \triangleright_2 \in B. \tag{7.10}$$

In particular, the left-hand side exists if and only if the right-hand side exists.

Identity *The set B includes an 'identity element' $\triangleright_e$ such that $a \triangleright_e b = a$ for all $a, b \in Q$.*

By abuse of notation, we often write Q as a shorthand for (Q, B).

Similar self-distributive structures have been studied in knot theory (*e.g.* (2)) and in the theory of computation (*e.g.* (12)). Specifically, if we were to

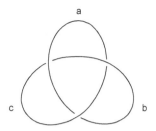

Figure 7.4: A knot diagram divided up into arcs.

require that all binary operations in B be fully defined—*i.e.* that $a \triangleright b$ exists for all $a, b \in Q$ and for all $\triangleright \in B$—then our notion would recover the notion of a multi-quandle (11).

7.3.2 Knot theoretic inspiration

A *knot* is a smooth embedding $K \colon S^1 \hookrightarrow \mathbb{R}^3$ of a directed circle in 3–space. A *knot diagram* is a projection of $K(S^1)$ onto a disjoint plane, together with over-under information at crossings. For example, the knot $K \stackrel{\text{def}}{=} (\sin t + 2\sin 2t, \cos t - 2\cos t, -\sin 3t)$ has the diagram of Figure 7.4 with respect to the plane $z = -2$. The *Reidemeister Theorem* in knot theory implies that two diagrams represent the same knot if and only if they differ by a finite set of *Reidemeister moves*, illustrated in Figure 7.5.

Given two knot diagrams, how can we show that they represent different knots? One method is to colour arcs of knot diagrams by formal variables $x, y, z, \ldots$ together with an operation $\triangleright$ defined by the condition that, for an overarc labeled y and an underarc labeled x, the remaining underarc in a crossing is labeled $x \triangleright y$. This defined a structure called the *fundamental quandle* of the diagram. Which underarc is labeled x and which $x \triangleright y$ is determined by the orientation of the overarc, as an oriented arc has a 'left' and a 'right'. A *quandle* is a set of formal variables together with an operation $\triangleright$ satisfying the axioms of Section 7.3.1. Each of these axioms corresponds to a Reidemeister move, and it is a theorem in knot theory that two diagrams represent the same knot if and only if they share the same fundamental quandle.

7.3.3 Linear and log-linear quandloids

From Equation 7.3, we see that an example of a quandloid is the set Q of all pairs $(\hat{X}, C)$ of an estimator $\hat{X}$ for a random variable X with an estimator C for its error covariance matrix, whose update operations are $\triangleright_s$ with $s \in [0, 1)$

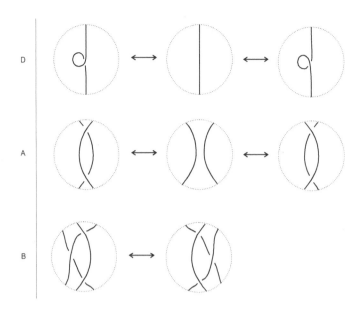

Figure 7.5: Reidemeister moves. To execute a Reidemeister move, cut out a disc inside a knot diagram containing one of the patterns above, and replace it with another disc containing the pattern on the other side of the Reidemeister move.

and their formal inverses $\lhd_s$ (where defined). Generalizing this example, we make a definition.

Definition 7.2 (Linear quandloid). *Equipping a real vector space $\bar{Q}$ with operations $\bar{\rhd}_s$ such that:*

$$a \,\bar{\rhd}_s\, b \stackrel{\text{def}}{=} (1-s)a + sb, \qquad a, b \in \bar{Q},$$

with $s \in S \subset \mathbb{R}$ where $1 \notin S$ (by causal invertibility) but $0 \in S$ (by identity) defines a quandloid called a linear *quandloid.*

A second class of examples of quandloids is given below:

Definition 7.3 (Log-linear quandloid). *Let Q be a space of measures defined over the σ-algebra of some set R, whose elements we think of as unnormalized probability density functions. Then equipping Q with the operations*

$$p(x) \rhd_s q(x) \stackrel{\text{def}}{=} p(x)^{1-s} q(x)^s, \qquad p(x), q(x) \in Q,$$

again with $s \in S \subset \mathbb{R}$ where $1 \notin S$ (by causal invertibility) but $0 \in S$ (by identity) defines a quandloid called a log-linear *quandloid.*

The log-linear quandloid has a geometric interpretation. The normalized counterpart of $p \rhd_s q$ can be viewed as a "point" along the "shortest information path" from $p(x)$ to $q(x)$, where the flow parameter $s \in [0,1)$, as may be verified by direct computation.

Our next quandloids will be derived from log-linear quandloids using *homomorphisms*.

Definition 7.4. *A* homomorphism *of quandloids is a function* $\boldsymbol{h} \stackrel{\text{def}}{=}$ $(h_E, h_V)\colon (Q, B) \to (\bar{Q}, \bar{B})$, *such that*

$$h_E(a \triangleright b) = h_E(a)\, h_V(\triangleright)\, h_E(b) \qquad \forall a, b \in Q, \quad \forall \triangleright \in B.$$

7.3.4 Covariance intersection

The covariance intersection update rules are obtained via a homomorphism from a particular log-linear quandloid. Explicitly:

Proposition 7.5. *Let* $Q \stackrel{\text{def}}{=} \Big\{ \exp\left(-\frac{1}{2}(X - \hat{X})^T C_X^{-1}(X - \hat{X})\right) \Big| \hat{X} \in \mathbb{R}^n$, $C_X \in \mathbb{R}^{n \times n} \Big\}$ *be the space of unnormalized Gaussian probability density functions over* $\mathbb{R}^n$. *Then the underlying space of parameters* $\bar{Q} = (\mathbb{R}^n, \mathbb{R}^{n \times n})$ *together with the operations* $\bar{\triangleright}_s \colon \bar{Q} \times \bar{Q} \to \bar{Q}$

$$\hat{Z} = \left(C_Z C_X^{-1} \hat{X}\right) \bar{\triangleright}_s \left(C_Z C_Y^{-1} \hat{Y}\right) = (1-s) C_Z C_X^{-1} \hat{X} + s C_Z C_Y^{-1} \hat{Y} \quad (7.11\text{a})$$

$$C_Z^{-1} = C_X^{-1} \bar{\triangleright}_s C_Y^{-1} = (1-s) C_X^{-1} + s C_Y^{-1} \tag{7.11b}$$

for every $(\hat{X}, C_X), (\hat{Y}, C_Y) \in \bar{Q}$, *with* $s \in [0,1)$, *form the linear quandloid of estimators and CI, considered at the beginning of this note.*

Proof. Let $p(X), p(Y) \in Q$ be unnormalized Gaussian densities parameterized by their mean and covariance, $(\hat{X}, C_X)$ and $(\hat{Y}, C_Y)$, respectively. An update operation reads:

$$p(X) \triangleright_s p(Y) = p(Z) =$$
$$= \exp\left(-\frac{1}{2}\left[(1-s)(Z - \hat{X})^T C_X^{-1}(Z - \hat{X}) + s(Z - \hat{Y})^T C_Y^{-1}(Z - \hat{Y})\right]\right)$$
$$\tag{7.12}$$

We verify that this expression equals:

$$\exp\left(-\frac{1}{2}(Z - \hat{Z})^T C_Z^{-1}(Z - \hat{Z})\right), \tag{7.13}$$

for $\hat{Z}$ the form which appears in 7.11a and for C_Z of the form which appears in 7.11b.

Given $p(X) \in Q$, we compute that $C_X^{-1} = -\frac{\partial^2 \log p(x)}{\partial x \partial x^T}$, which is called the *precision matrix* of $p(X)$. In particular,

$$C_Z^{-1} = -\frac{\partial^2 \log p(z)}{\partial z \partial z^T} = (1-s)C_x^{-1} + sC_Y^{-1} = C_x^{-1} \bar{\triangleright}_s C_Y^{-1} \qquad (7.14)$$

which follows from (7.12).

For any $p(X) \in Q$ the mode $\hat{X}$ satisfies:

$$\left.\frac{\partial \log p(x)}{\partial x}\right|_{x=\hat{X}} = 0 \qquad (7.15)$$

The mode $\hat{Z}$ of $p(Z)$ is thus obtained from (7.12) by

$$2\left.\frac{\partial \log p(z)}{\partial z}\right|_{z=\hat{Z}} = (1-s)C_x^{-1}(\hat{Z} - \hat{X}) + sC_Y^{-1}(\hat{Z} - \hat{Y}) = 0 \qquad (7.16)$$

finally yielding

$$C_z^{-1}\hat{Z} = (1-s)C_x^{-1}\hat{X} + sC_Y^{-1}\hat{Y} = \left(C_x^{-1}\hat{X}\right) \bar{\triangleright}_s \left(C_Y^{-1}\hat{Y}\right). \qquad (7.17)$$

$\square$

7.3.5   Fisher information

Let I_θ be the *observed* Fisher information matrix associated with the unnormalized joint probability density function:

$$(p \triangleright_s q)(X, \theta) \stackrel{\text{def}}{=} p(X, \theta) \triangleright_s q(X, \theta)$$

where θ is a random parameter vector whose prior is $g(\theta)$, and where $p(X, \theta) = p(X \mid \theta)g(\theta)$, $q(X, \theta) = q(X \mid \theta)g(\theta)$. Then I_θ satisfies:

$$\underbrace{-\frac{\partial^2 \log(p \triangleright_s q)(X, \theta)}{\partial \theta_i \partial \theta_j}}_{I_\theta(p \triangleright_s q)} = \underbrace{\left(-\frac{\partial^2 \log p(X, \theta)}{\partial \theta_i \partial \theta_j}\right)}_{I_\theta(p)} \bar{\triangleright}_s \underbrace{\left(-\frac{\partial^2 \log q(X, \theta)}{\partial \theta_i \partial \theta_j}\right)}_{I_\theta(q)} \qquad (7.18)$$

where:

$$a\,\bar{\triangleright}_s b \stackrel{\text{def}}{=} (1-s)a + sb, \qquad a, b \in Q,$$

is a binary operation for a linear quandloid. Note that, in the case of a random parameter vector, the normalizing factor is independent of the parameters, and hence we may use the unnormalized $p\triangleright_s q$ in the expression for the observed Fisher information matrix.

Note that *expected* Fisher information matrices do not form a quandloid in general. Exceptional cases, when expected Fisher information matrices do form a quandloid, are either when $I_\theta = E\{I_\theta\}$ or when the underlying expectation is taken exclusively with respect to the prior $g(\theta)$. We give examples of both of these cases.

Assume that $p(X, \theta) = p(X \mid \theta)g(\theta)$ and $q(X, \theta) = q(X \mid \theta)g(\theta)$ are colours of a log-linear Gaussian quandloid Q, where θ is a mode whose prior $g(\theta)$ is also Gaussian. Then $h_E(\cdot) \overset{\text{def}}{=} I_\theta(\cdot)$ is a homomorphism and $(E\{\mathcal{I}_\theta\}, \bar{\rhd}_s)$ is a linear quandloid:

$$E\{I_\theta(p \rhd_s q)\} = (C_p^{-1} + C_g^{-1})\bar{\rhd}_s(C_q^{-1} + C_g^{-1}) = E\{I_\theta(p)\}\bar{\rhd}_s E\{I_\theta(q)\}$$

where C_p, C_q and C_g denote the covariances of p, q and g, and $s \in [0, 1)$.

Here is another example. Consider a log-linear quandloid whose colours are of the form $p(X, \theta) = p(X \mid \lambda_p(\theta))g(\theta)$ and $q(X, \theta) = q(X \mid \lambda_q(\theta))g(\theta)$ where the underlying conditionals are exponential densities whose θ-dependent rate parameters are λ_p and λ_q. As before $g(\theta)$ denotes the prior of θ. Additionally, we assume that the second derivative of any rate parameter vanishes $d^2\lambda(\theta)/d\theta^2 = 0$. In such a case, indeed:

$$E\{I_\theta(p \rhd_s q)\} = E\{I_\theta(p)\}\bar{\rhd}_s E\{I_\theta(q)\}$$

and thus $(E\{\mathcal{I}_\theta\}, \{\bar{\rhd}_s\})$ a linear quandloid.

7.3.6 Shannon information

A binary information source is described by a Bernoulli random variable X. The entropy of X, denoted $H(X)$, has an operational meaning given by Shannon's source coding theorem. Very long independent identically distributed sequences generated by such a source fall into two categories: they are either *typical* or not. The probability of a typical sequence stabilizes around the value $2^{-NH(X)}$ which implies that not more than $NH(X)$ bits are required to encode any typical message with 'negligible' loss of information.

Let Q be a log-linear quandloid whose colours are uniform probability densities of typical sequences associated with infinitely many information sources

$$Q \overset{\text{def}}{=} \left\{ 2^{-NH(X)} \mid X \quad \text{is an information source} \right\}$$

The set of entropies $\{H(X)\}$ with operations $\bar{\rhd}_s$ constitute a linear quandloid:

$$H(Z) = H(X)\bar{\rhd}_s H(Y) = (1-s)H(X) + sH(Y), \qquad H(X), H(Y) \in \mathcal{H}$$

which follows from the homomorphism $h_E(\cdot) \overset{\text{def}}{=} -\frac{1}{N}\log(\cdot)$.

A source Z whose entropy is obtained as above behaves sometimes as X and some other times as Y. For example, generating a lengthy message of

N symbols from Z one expects that approximately $(1 - s)N$ of them will make up a typical sequence from X whereas the rest sN symbols will look as though they where generated by Y.

An update in a fusion tangle coloured by $(\mathcal{H}, \bar{\triangleright}_s)$ may be interpreted in one of two ways. It may either describe a source Z which behaves like X with probability $(1 - s)$ and like Y otherwise. For example, generating a message of N symbols from Z, one expects that approximately $(1 - s)N$ of them will make up a typical sequence from X whereas the remaining sN symbols will look as though they where generated by Y. Alternatively, an update may describe a concatenation of messages from the two input sources, *i.e.* $NH(Z)$ is the number of bits required to encode a concatenation of length $(1 - s)N$ and sN messages generated by, respectively, X and Y.

7.4   What Is an Information Fusion Tangle?

We assemble information updates into a network. The philosophical position underlying its definition is that non-associative self-distributive algebraic structures (*e.g.* quandloids) should not label graphs, but rather they should label low-dimensional 'tangled' topological objects.

Definition 7.6 (Information fusion tangle). *An* information fusion tangle *is:*

▪ *A directed graph G whose vertices (drawn as boxes e.g. in Figure 7.3) have either degree 1 or have positive even degree. Vertices of even degree are called* interactions, *and the in-degree of each interaction equals its out-degree. Vertices of degree 1 are called* endpoints. *An endpoint is* initial *if it is a source, and is* terminal *if it is a sink.*

▪ *For each interaction v, whose degree we denote $2m+2$, a partition of edges incident to v into pairs:*

$$\left\{ (f^{\ in}, f^{\ out}) \right\} \cup \left\{ \begin{array}{l} (e_1^{\ in}, e_1^{\ out}), (e_2^{\ in}, e_2^{\ out}), \dots \\ (e_m^{\ in}, e_m^{\ out}) \end{array} \right\}.$$

We consider each of the pairs $(f^{\ in}, f^{\ out})$ as though it were a single edge, and refer to that pair as the agent *of v, and we call each $e_i^{\ in}$ an input *of v, and its corresponding edge $e_i^{\ out}$ an* output *of v. All edges with superscript 'in' (i.e. $f^{\ in}$ and $e_1^{\ in}, e_1^{\ in}, \dots, e_m^{\ in}$) are directed towards v, and all edges with superscript 'out' are directed away from v.*

▪ *A colouring function $\boldsymbol{\rho} \overset{\text{def}}{=} (\rho_V, \rho_E)$ with $\rho_V \colon V(G)_{\deg \geq 2} \to B$ and $\rho_E \colon E(G) \to Q$, where (Q, B) is a quandloid, satisfying:*

1. *If* (f^{in}, f^{out}) *is the agent of* v, *then* $\rho_E(f^{in}) = \rho_E(f^{out})$.

2. *For each input-output pair* (e_i^{in}, e_i^{out}) *we have either* $\rho_E(e_i^{out}) = \rho_E(e_i^{in}) \rhd \rho_E(f)$ *or* $\rho_E(e_i^{in}) = \rho_E(e_i^{out}) \rhd \rho_E(f)$ *where* $\rhd$ *denotes* $\rho_V(v)$, *and* $\rho_E(f) = \rho_E(f^{in}) = \rho_E(f^{out})$ *is the colour of the agent of* v.

Each transition $\rho_E(e_i^{in}) \mapsto \rho_E(e_i^{out})$ *is called an* update *if* $\rho_E(e_i^{out}) = \rho_E(e_i^{in}) \rhd \rho_E(f)$ *and is called a* discount *if* $\rho_E(e_i^{in}) = \rho_E(e_i^{out}) \rhd \rho_E(f)$.

Our diagrammatic convention is to update from right to left of the agent edge(s), so that if the agent edge is drawn pointing from top to bottom then the colour of an edge to the left of a box is updated by $\rhd$ to become the colour of the corresponding edge the right of the box, and if the agent edge is drawn pointing from bottom to top then the opposite.

Note that the number of initial endpoints of an information fusion tangle equals its number of terminal endpoints. This relates information fusion to reversible computation, thanks to causal invertibility (6).

7.5 Information Fusion Tangle Equivalence

In this section we describe three local modifications on information fusion tangles. Two information fusion tangles which differ by a finite sequence of these modifications are considered to be *equivalent*. For this reason, we think of these modifications as *conservation laws*.

Reidemeister I   This local move modifies a tangle by eliminating or introducing an update of a edge by itself. By coherence of information, no information is gained or lost when we perform this move.

Reidemeister II   This local move modifies a tangle by updating an edge, and immediately discounting it by the same agent. By causal invertibility doing such a thing does not effect the colour of the edge, and so no information is gained or lost when we perform this move.

Reidemeister III   This local move modifies a tangle by replacing updated outputs with a common agent by corresponding updated inputs updated by the correspondingly updated agent. By 'no double counting' doing such a thing does not effect the colour of the edge, and so no information is gained or lost when we perform this move.

In low dimensional topology, the *Reidemeister Theorem* tells us that

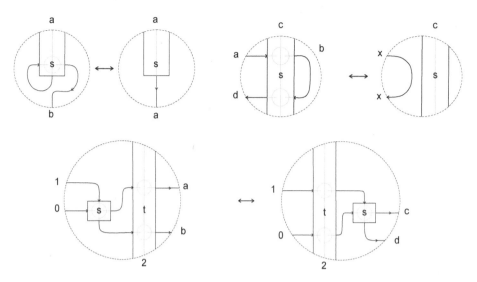

Figure 7.6: The Reidemeister Moves, which are local modifications of information fusion tangles. The moves are considered for all orientations on all edges. Only a special case of R3 is illustrated above— the general case is given by Figure 7.7.

Figure 7.7: For simplicity, we draw boxes by thickening the agent. We can now easily draw the R3 move.

two diagrams of tangles are related by a finite sequence of Reidemeister moves if and only if the tangles which they represent are *ambient isotopic*. Combined with a result stating that a tangle is equivalent to the set of all of its colourings, tells us that the Reidemeister moves are the unique set of local moves on tangles which preserve the information content of the tangle— *i.e.* the set of all possible tangle colourings. The appearance of Reidemeister moves signifies that the theory of information fusion tangles has a low dimensional topological flavour.

We believe that the Reidemeister moves are diagrammatic representations of fundamental symmetries of information fusion (although there may also be other symmetries), and therefore that the formalism of information fusion tangles subject to Reidemeister moves is a topological candidate for a suitably flexible diagrammatic language with which to discuss networks of information fusion.

7.6   A More Sophisticated Example

In this section, we present a more sophisticated version of the example in Section 7.2.

Consider the information fusion tangle illustrated in the upper left corner of Figure 7.8. This tangle has a set of outputs outside the bounding disk (not shown). Some set of intermediate edges lies inside the sub-tangle designated by N_L and graphically represented by an empty circle.

Erroneous data cause one or more of the edges 0,1, and 2, to carry faulty pieces of information, *e.g.* biased, inconsistent and otherwise unreliable estimates. We may bring the network to its optimal configuration by 'sliding' the faulty edges all the way over N_L, by repeated application of the second and third Reidemeister moves. In the resulting topology, the faulty edges have no effect on N_L, and so local costs are improved.

Acknowledgment

This research was supported by the Israel Science Foundation (grant No. 1723/16).

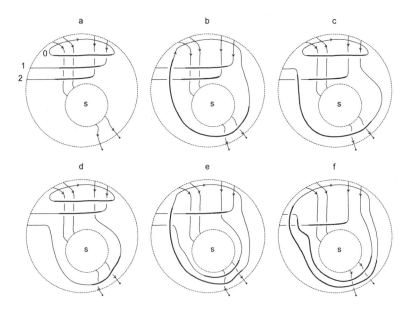

Figure 7.8: The fault code appearing above each network should be read as follows. The digits from left to right correspond to edges 2,1 and 0. An "X" in the place of the *i*th digit signifies faulty content in its respective edge.

References

1. Bass, T. 2000 Intrusion detection systems and multisensor data fusion. *Communications of the ACM* **43**(4), 99–105.

2. Carter, S. J. 2009 A survey of quandle ideas. In *Introductory lectures on Knot Theory: Selected lectures presented at the Advanced School and Conference on Knot Theory and its Applications to Physics and Biology*, 22–53.

3. Hall, D. L. Llinas, J. 1997 An introduction to multisensor data fusion. *Proceedings of the IEEE* **85**(1), 6–23.

4. Julier, S. Uhlmann, J. K. 2001 General decentralized data fusion with covariance intersection. In *Handbook of Multisensor Data Fusion*, CRC Press, Boca Raton, FL.

5. Khaleghi, B. Khamis, A. Karray, F. O. Razavi, S. N. 2013 Multisensor data fusion: A review of the state-of-the-art. *Information Fusion* **14**(1), 28–44.

6. Landauer, R. 1961, Irreversibility and heat generation in the computing process. *IBM Journal of Research and Development*, vol. 5, 183–191.

7. Lou, R. C. Chih-Chen, Y. Kuo, L. S. 2002 Multisensor fusion and

integration: approaches, applications, and future research directions. *IEEE Sensors Journal* **2**(2), 107–119.

8. Murphy, R. 2000 Introduction to AI robotics. *MIT Press*

9. Nakamura, E. F. Loureiro, A. F. Frery, A. C. 2007 Information fusion for wireless sensor networks: Methods, models, and classifications. *ACM Computing Surveys* **39**(3)

10. Nielsen, F. 2013 An Information-Geometric Characterization of Chernoff Information. *IEEE Signal Processing Letters* **20**(3), 269–272.

11. Przytycki, J. H. 2011 Distributivity versus associativity in the homology theory of algebraic structures. *Demonstration Mathematica* **44**(4), 823–869. `arXiv:1109.4850`

12. Roscoe, A. W. 1990 Consistency in distributed databases. *Oxford University Computing Laboratory Technical Monograph* **PRG-87**.

13. Ross, A. Jain, A. 2003 Information fusion in biometrics. *Pattern Recognition Letters* **24**(13), 2115–2125.

14. Uhlmann, J. K. 2003 Covariance consistency methods for fault-tolerant distributed data fusion. *Information Fusion* **4**, 201–215.

15. Xiong, N. Svensson, P. 2002 Multi-sensor management for information fusion: issues and approaches. *Information Fusion* **3**, 163–186.

16. Yong, D. WenKang, S. ZhenFu, Z. Qi, L. 2004 Combining belief functions based on distance of evidence. *Decision Support Systems* **38**(3), 489–493.

Optimization Techniques to Improve Training Speed of Deep Neural Networks for Large Speech Tasks

Tara N. Sainath   tsainath@google.com
Google Research
New York, NY, USA

Brian Kingsbury   bedk@us.ibm.com
Thomas J. Watson Research Center
Yorktown Heights, NY, USA

Hagen Soltau   soltau@google.com
Thomas J. Watson Research Center
Yorktown Heights, NY, USA

Bhuvana Ramabhadran   bhuvana@us.ibm.com
Thomas J. Watson Research Center
Yorktown Heights, NY, USA

While Deep Neural Networks (DNNs) have achieved tremendous success for large vocabulary continuous speech recognition (LVCSR) tasks, training these networks is slow. Even to date, the most common approach to train DNNs is via stochastic gradient descent, serially on one machine. Serial training, coupled with the large number of training parameters (i.e., 10-50 million) and speech data set sizes (i.e., 20-100 million training points) makes DNN training very slow for LVCSR tasks. In this work, we explore a variety of different optimization techniques to improve DNN training speed. This includes parallelization of the gradient computation during cross-entropy and sequence training, as well as reducing the number of parameters in the network using a low-rank matrix factorization. Applying the proposed optimization techniques, we show that DNN training can be sped up by a factor of 3 on a 50-hour English Broadcast News (BN) task with no loss in accuracy. Furthermore, using the proposed techniques, we are able to train

DNNs on a 300-hr Switchboard (SWB) task and a 400-hr English BN task, showing improvements between 9-30% relative compared to a state-of-the art GMM/HMM system while the number of parameters of the DNN is smaller than the GMM/HMM system.

8.1   Introduction

Deep Neural Networks (DNNs) have become a popular acoustic modeling technique in the speech community over the last few years (7), showing significant gains over state-of-the-art Gaussian Mixture Model/Hidden Markov Model (GMM/HMM) systems on a wide variety of small and large vocabulary tasks. The development of pre-training algorithms (9) and better forms of random initialization (6), as well as the availability of faster computers, has made it possible to train deeper networks than before, and in practice these deep networks have achieved excellent performance (10, 12, 30).

However, one drawback of DNNs is that training remains very slow, particularly for large vocabulary continuous speech recognition (LVCSR) tasks. This can be attributed to a variety of reasons. First, models for real-world speech tasks are trained on hundreds of hours of data, which amounts to many millions of training examples. It was shown in (35), that DNN performance improves with increasing training data. Second, roughly 10-50 million DNN parameters are used for speech tasks (25), (30), which is much larger compared to the number of parameters used with common acoustic modeling approaches for speech recognition (i.e., Gaussian Mixture Models (GMMs)) on the same tasks. Third, to date the most popular methodology to train DNNs is with stochastic gradient descent (SGD), serially on one machine. The objective of this chapter is to address latter two problems with DNN training, namely the large number of parameters and serial training.

First, we explore improving training time of cross-entropy backpropagation by parallelizing the gradient computation. During SGD training, the gradient is computed over a small collection of training examples, known as a mini-batch. Typically, gradient parallel SGD methods are not effective in speech tasks because the size of this mini-batch is small (i.e., 128-512) (30) and the number of DNN parameters is large (i.e., 10-50 million). Therefore, there is a large communication cost involved computing gradients on subsets of this mini-batch on each worker, and passing these large gradient vectors back to a master (15), (33). In this paper, we explore a hybrid pre-training strategy (13) that introduces an objective function which combines the generative benefits of unsupervised pre-training (9) and the discriminative benefits of the supervised cross-entropy objective function. We will show that a benefit of hybrid pre-training is that the weights are in a much better initial space relative to generative pre-training. This allows the mini-batch size during fine-tuning to be made very large relative to generative pre-training, and thus the gradient can be parallelized effectively to improve overall training speed.

Second, we explore parallelizing the gradient computation during sequence training. Sequence-training is often performed after cross-entropy (CE) training and readjusts the CE weights using a sequence-level objective function. Sequence training usually provides an additional 10-15% relative improvement in word error rate (WER) on top of CE training (12). While parallel SGD can be used to improve training time for cross-entropy, sequence-training involves loading large lattice files which carry sequence-level information, which requires significant bandwidth. Therefore, parallelizing across few machines (i.e., 4-5) as is done with parallel SGD is not enough machines to allow for a large improvement in speed. Naturally, adding more workers increases communication costs and does not lead to speed improvements either. Hessian-free (HF) sequence training (18) has been proposed as an alternative to SGD. One of the benefits of HF training is that the gradient is computed on all of the data instead of a large batch, and thus lends itself more naturally to parallelization across more machines (i.e. 40-80).

Third, we explore reducing parameters of the DNN before training, such that overall training time is reduced. Typically in speech, DNNs are trained with a large number of output targets (i.e., 2,000–10,000), equal to the number of context-dependent states of a GMM/HMM system, to achieve good recognition performance. Having a larger number of output targets contributes significantly to the large number of parameters in the system, as over 50% of parameters in the network can be contained in the final layer. Furthermore, few output targets are actually active for a given input, and we hypothesize that the output targets that are active are probably correlated (i.e. correspond to a set of confusable context-dependent HMM states). The last weight layer in the DNN is used to project the final hidden representation to these output targets. Because few output targets are active, we suspect that the last weight layer (i.e., matrix) has low rank. If the matrix is low-rank, we can use factorization to represent this matrix by two smaller matrices, thereby significantly reducing the number of parameters in the network before training. Another benefit of low-rank factorization for non-convex objective functions, such as those used in DNN training, is that it constrains the space of search directions that can be explored to maximize the objective function. This helps to make the optimization more efficient and reduce the number of training iterations, particularly for 2nd-order optimization techniques.

Our initial experiments are conducted on a 50-hour English Broadcast News (BN) task (11). We show that with hybrid pre-training + parallel SGD, we can achieve roughly a 2.5 times speedup in fine-tuning time over generative pre-training and a small batch size, with a very small decrease in accuracy. Furthermore, including low-rank factorization, we can achieve a 3 times speedup over the generative pre-training + small batch size. Second, we explore the speedups obtained with HF for sequence-training, showing that we can achieve roughly a 3 times speedup over SGD for sequence training. Furthermore, including low-rank factorization into sequence-training gives a 4 times speedup. We then explore the scalability of the three optimization methods (i.e., parallel SGD, HF and low-rank) to two larger tasks, namely a 300-hour Switchboard and a 400-hour BN task. With these techniques, we are still able to achieve between a 10-20% relative improvement over a state-of-the art GMM/HMM baseline, consistent with the gains observed on similar tasks in the literature (10), (12), (30).

The rest of this paper is organized as follows. Related work on improving training time for DNNs is presented in Section 8.2. Section 8.3 discusses hybrid pre-training and parallel SGD for cross-entropy training, while Section 8.4 outlines the HF algorithm for sequence training. Low-rank matrix factorization is explained in Section 8.5. Experiments and results on the three proposed optimization techniques are presented for a 50-hour BN task in Section 8.6 and for larger tasks in Section 8.7. Finally, Section 8.8 concludes the paper and discusses future work.

8.2 Related Work

In this section, we present a literature survey of past work explored to improve DNN training speed.

8.2.1 Parallel Methods

Stochastic Gradient Descent (SGD) remains one of the most popular approaches to training DNNs. SGD methods are simple to implement and are generally faster for large data sets compared to 2nd order methods (15). While parallel SGD methods have been successfully explored for convex problems (38), for non-convex problems such as DNNs, it is very difficult to parallelize SGD across machines. With SGD the gradient is computed over a small collection of frames (known as a mini-batch), which is typically on the order of 100-1,000 for speech tasks (8). Splitting this gradient computation onto a few parallel machines, coupled with the large number of network parameters used in speech tasks, results in large communications costs in passing the gradient vectors from worker machines back to the master. Thus, it is generally cheaper to compute the gradient serially on one machine using the standard SGD technique (27). It is important to note that recently (4) explored a distributed asynchronous SGD method to improve DNN training speed.

Batch methods, including conjugate gradient (CG) or limited-memory BFGS (L-BFGS), generally compute the gradient over all of the data rather than a mini-batch, and therefore are much easier to parallelize (3). However, as shown in (15), parallelization of dense networks can actually be slower than serial SGD training, again because of communication costs in passing models and gradients as well as the need to run more training iterations compared to serial SGD. Therefore, parallelization methods for DNNs have not enjoyed much success.

8.2.2 Reducing Parameters

There have been a few attempts in the speech recognition community to reduce number of parameters in the DNN. One common approach, known as "optimal brain damage" (17), uses a curvature measure to decide which weights to zero out. In addition, a "sparsification" approach proposed in (30) looks to implicity zero out weights which are below a certain threshold (i.e. close to zero). Both of these techniques are meant to act as a regularizer and improve generalization of the network. However, it reduces parameters after the network architecture has been defined, and therefore does not have any impact on training time, though it can be used to improve decoding time (36).

Second, convolutional neural networks (CNNs) (16) have also been explored, primarily in computer vision, to reduce parameters of the network by sharing weights across both time and frequency dimensions of the speech signal. However, experiments show that in speech recognition, the best performance with CNNs is achieved when matching the number of parameters to a DNN (23), and therefore parameter reduction with CNNs does not always hold in speech tasks.

8.2.3   Low-Rank Factorization

The use of low-rank matrix factorization for improving optimization problems has been explored in a variety of contexts. For example, in multivariate regression involving a large-number of target variables, the low-rank assumption on model parameters has been effectively used to constrain and regularize the model space (37) leading to superior generalization performance. DNN training may be viewed as effectively performing nonlinear multivariate regression in the final layer, given the data representation induced by the previous layers. Furthermore, low-rank matrix factorization algorithms also find wide applicability in matrix completion literature (see, e.g., (22) and references therein). Our work extends these previous works by exploring low-rank factorization specifically for DNN training, which has the benefit of reducing the overall number of network parameters and improving training speed.

8.2.4   Improved Hardware

Graphical processors (GPUs) have become a popular hardware solution to speed up DNN training compared to multi-core CPUs (21). GPUs have hundreds of cores compared to multi-core CPUs, and can parallelize the matrix-multiplication during DNN training quite efficiently. This can allow for greater than a 5x speedup in training time compared to CPUs. However, one problem with GPUs is that they are expensive relative to multi-core CPUs, and thus many computing infrastructures contain many more CPUs compared to GPUs. In this paper, we specifically focus on improving DNN training time for CPUs.

8.3   Parallel Stochastic Gradient Descent

In this section, we describe a hybrid pre-training strategy that allows for cross-entropy fine-tuning to be parallelized.

8.3.1   Pre-Training Strategies

8.3.1.1   *Generative Pre-Training*

The Restricted Boltzmann Machine (RBM) is a commonly used model for generative pretraining (9). An RBM is a bipartite graph where visible units $\mathbf{v}$, representing observations, are connected via undirected weights to hidden units **h**. Units **h** and **v** are stochastic, with values distributed according to a given distribution, and the entire RBM is endowed with an energy function. For an RBM in which all units are binary, and follow a Bernoulli distribution, the energy function is

$$E(\mathbf{v}, \mathbf{h}; \theta) = -\mathbf{h}^T \mathbf{W} \mathbf{h} - \mathbf{b}^T \mathbf{v} - \mathbf{a}^T \mathbf{h} \tag{8.1}$$

where $\theta = \{\mathbf{W}, \mathbf{b}, \mathbf{a}\}$ defines the RBM parameters, including weights **h**, visible biases **b**, and hidden biases **a**.

The RBM assigns a probability to an observed vector **v** based on the energy function

$$p(\mathbf{v}; \theta) = \frac{\sum_{\mathbf{h}} e^{-E(\mathbf{v}, \mathbf{h}; \theta)}}{\sum_{\mathbf{u}} \sum_{\mathbf{h}} e^{-E(\mathbf{u}, \mathbf{h}; \theta)}} \tag{8.2}$$

and the RBM parameters are trained to maximize this generative likelihood.

In generative pre-training, an RBM is used to learn the weights for the first layer of a neural network. Once these weights are learned, the outputs (hidden units) are treated as inputs to another RBM that learns higher-order features, and the process is iterated for each layer in the network. Because speech features are continuous, the RBM for the first layer is a Gaussian-Bernoulli RBM. Subsequent layers are trained using Bernoulli-Bernoulli RBMs. This greedy, layer-wise pre-training scheme is both fast and effective (9). After a stack of RBMs has been trained, the layers are connected together to form what is referred to as a DNN.

8.3.1.2 *Discriminative Pre-Training*

Rather than maximizing the generative likelihood $p(\mathbf{v}; \theta)$ as in generative pre-training, discriminative pre-training optimizes the likelihood $p(\mathbf{l}|\mathbf{v}; \theta)$, which makes use of both features **v** and labels **l** (14), (29). This discriminative likelihood is defined to be the cross-entropy objective function which is used during fine-tuning (i.e. backpropagation). Training an RBM discriminatively is referred to as DRBM (13).

In the discriminative pre-training methodology, a 2-layer DRBM, namely one hidden layer and one softmax layer, is trained using the cross-entropy criteria with label information. After taking one pass through the entire data with discriminative pre-training, the softmax layer is thrown away and replaced by another randomly initialized hidden layer and softmax layer on top. The initially trained hidden layer is held constant, and discriminative pre-training is performed on the new hidden and softmax layers. This discriminative training is greedy and layer-wise like generative RBM pre-training.

8.3.1.3 Hybrid Pre-Training

One problem with performing discriminative pre-training is that at every layer, weights are learned to minimize the objective function (i.e., cross-entropy). This means that weights learned in lower layers are potentially not general enough, but rather too specific to the final DNN objective. Having generalized weights in lower layers has been shown to be helpful. Specifically, generalized concepts, such as mapping phones from different speakers into a canonical space, are captured in lower layers, while more discriminative representations such as differentiating between different phonemes, are captured in higher layers (19).

Hybrid pre-training has been proposed to address the issues of discriminative pre-training, by performing pre-training with both a generative and discriminative component. We follow a hybrid pre-training recipe similar to the methodology in (13), which looks to maximize the objective function in Equation 8.3, where α is an interpolation weight between the discriminative $p(\mathbf{l}|\mathbf{v})$ and generative components $p(\mathbf{l}, \mathbf{v})$. More intuitively, the generative component can be seen to act as a data-dependent regularizer for the discriminative component (13). The hybrid discriminative methodology is referred to as HDRBM. While (13) only explored pre-training a two layer HRDBM with binary inputs, in this work we extend to multiple layers and continuous inputs.

$$p(\mathbf{l}|\mathbf{v}) + \alpha p(\mathbf{v}, \mathbf{l}) \tag{8.3}$$

To optimize the generative component $p(\mathbf{v}, \mathbf{l})$, first consider a 2 layer DNN, where the weights, hidden and visible biases for layer 1 are given by $\{W, \mathbf{a}, \mathbf{b}\}$, and for layer 2 as $\{U, \mathbf{c}, \mathbf{d}\}$. In addition we will define $\mathbf{l}$ to be a labels vector with an entry of 1 corresponding to the class label of input $\mathbf{v}$ and zeros elsewhere. For an HDRBM in which all units are binary and follow a Bernoulli distribution, the energy function is given by Equation 8.4:

$$E(\mathbf{v}, \mathbf{l}, \mathbf{h}; \theta) = -\mathbf{h}^T W \mathbf{v} - \mathbf{b}^T \mathbf{v} - \mathbf{a}^T \mathbf{h} - \mathbf{c}^T \mathbf{l} - \mathbf{h}^T U \mathbf{l} \tag{8.4}$$

The joint probability that the model assigns to a visible vector $\mathbf{v}$ and label $\mathbf{l}$ is given by Equation 8.5:

$$p(\mathbf{v}, \mathbf{l}; \theta) = \frac{\sum_{\mathbf{h}} e^{-E(\mathbf{v}, \mathbf{l}, \mathbf{h})}}{\sum_{\mathbf{u}} \sum_{\mathbf{k}} \sum_{\mathbf{h}} e^{-E(\mathbf{u}, \mathbf{k}, \mathbf{h})}} \tag{8.5}$$

The generative component is trained to maximize the likelihood $p(\mathbf{v}, \mathbf{l}; \theta)$, while the discriminative component $p(\mathbf{l}; \theta)$ is trained similar to the discriminative pre-training methodology. To train an HDRBM, stochastic gradient descent is used, and for each example the gradient contribution due to $p(\mathbf{l}; \theta)$ is added to α times the gradient estimated from $p(\mathbf{v}, \mathbf{l}; \theta)$. Similar to RBM training, because input speech features are continuous, the HDRBM for the first layer is a Gaussian-Bernoulli HDRBM, while subsequent layers are Bernoulli-Bernoulli HDRBMs. Again, training is performed in a greedy, layerwise fashion similar to discriminative pre-training.

8.3.2 Stochastic Gradient Descent

During fine-tuning, each frame is labeled with a target class label. Given a DNN and a set of pre-trained weights, fine-tuning is performed via back-propagation to retrain the weights such that the loss between the target and hypothesized class probabilities is minimized. During SGD fine-tuning, the gradient is estimated using a small collection of frames, which is referred to as a mini-batch. (9). The weight update per mini-batch is given more explicitly by Equation 8.6, where γ is the learning rate, $\mathbf{w}$ are the weights, $\mathbf{v}_i$ is training example i and $\nabla_w f(\mathbf{w}; \mathbf{v}_i)$ is the gradient of the objective function $f(\mathbf{w}; \mathbf{v}_i)$ computed using this training example and weights. In addition, B is the mini-batch size. Examples of the objective function $f(\mathbf{w}; \mathbf{v}_i)$ commonly used include cross-entropy, minimum bayes risk and miminum mean-squared error.

$$\mathbf{w} := \mathbf{w} - \gamma \sum_{i=1}^{B} \nabla_w f(\mathbf{w}; \mathbf{v}_i) \tag{8.6}$$

Notice from Equation 8.6 that the gradient is calculated as the sum of gradients from individual training examples. When the batch size B is large (and thus number of training examples large), this allows the gradient computation to be parallelized across multiple worker computers. Specifically, on each worker a gradient is estimated using a subset of training examples, and then the gradients calculated on each slave computer are added together by a master computer to estimate the total gradient. We will observe that when using hybrid pre-training and having a much better initial weight space, the mini-batch size can be increased and the gradient computation can efficiently be parallelized.

8.4   Hessian-Free Optimization

In this section, we discuss speeding up sequence-training with Hessian-free optimization.

8.4.1   Motivation

In (11) it was shown that the lattice-based machinery developed for sequence-discriminative training of GMMs can be used for neural networks, and that the state-level minimum Bayes risk (sMBR) criterion improves word error rate by 18% relative over cross-entropy on a 50-hour English broadcast news task. However, one of the shortcomings of the experiments in (11) is that the networks were underparameterized for the amount of training data, using only 384 quinphone states and 153K weights. Both generative pretraining and discriminative cross-entropy training of a deep neural network using 9,300 triphone states and 45.1M parameters (16.1M non-zero parameters following sparsification) have been scaled to a 300-hour Switchboard task by using GPGPU hardware and caching training data in memory (30). However, even with high-performance hardware and careful algorithmic development, training still required about 30 days (29). Sequence-discriminative training is potentially even more expensive because the lattices required for the gradient computation are too large to cache in memory. This motivates exploration of distributed algorithms that split computation and I/O across multiple nodes in a compute cluster.

While parallelized SGD can be used to improve training time for cross-entropy, sequence-training involves loading in large lattice files, and thus parallelizing across 4-5 machines (i.e., workers) is not enough machines to allow for a large improvement in training speed. Naturally, adding more workers increases communication costs and does not lead to speed improvements either. Therefore, we seek a batch-method solution for sequence training. This allows the gradient to be computed on all the data and allows for increased number of workers, thus allowing lattices to more efficiently be loaded onto worker machines.

8.4.2   Algorithm

The challenge in performing distributed optimization is to find an algorithm that uses large data batches that can be split across compute nodes without incurring excessive overhead, but that still achieves performance competitive with stochastic gradient descent. One class of algorithms for this problem uses second-order optimization, with large batches for the gradient and much smaller batches for stochastic estimation of the curvature (18, 1, 34). A distributed implementation of one such algorithm has already been applied to learning an exponential model with a convex objective function for a speech recognition task (1).

The current study uses Hessian-free optimization (18) because it is specifically designed for the training of deep neural networks, which is a non-convex problem. Let $\boldsymbol{\theta}$ denote the network parameters, $\mathcal{L}(\boldsymbol{\theta})$ denote a loss function, $\nabla\mathcal{L}(\boldsymbol{\theta})$ denote the gradient of the loss with respect to the parameters, $\mathbf{d}$ denote a search direction, and $\mathbf{B}(\boldsymbol{\theta})$ denote a matrix characterizing the curvature of the loss around $\boldsymbol{\theta}$. The central idea in Hessian-free optimization is to iteratively form a quadratic approximation to the loss,

$$\mathcal{L}(\boldsymbol{\theta} + \mathbf{d}) \approx \mathcal{L}(\boldsymbol{\theta}) + \nabla\mathcal{L}(\boldsymbol{\theta})^T\mathbf{d} + \frac{1}{2}\mathbf{d}^T\mathbf{B}(\boldsymbol{\theta})\mathbf{d} \tag{8.7}$$

and to minimize this approximation using conjugate gradient (CG), which accesses the curvature matrix only through matrix-vector products $\mathbf{B}(\boldsymbol{\theta})\mathbf{d}$ that can be computed efficiently for neural networks (20). If $\mathbf{B}(\boldsymbol{\theta})$ were the Hessian and conjugate gradient were run to convergence, this would be a matrix-free Newton algorithm. In the Hessian-free algorithm, the conjugate gradient search is truncated, based on the relative improvement in approximate loss, and the curvature matrix is the Gauss-Newton matrix (28), which unlike the Hessian is guaranteed positive semidefinite, with additional damping: $\mathbf{G}(\boldsymbol{\theta}) + \lambda\mathbf{I}$.

Our implementation of Hessian-free optimization, which is illustrated as pseudocode in Algorithm 8.1, closely follows that of (18), except that it currently does not use a preconditioner. Gradients are computed over all the training data. Gauss-Newton matrix-vector products are computed over a sample (about 1% of the training data) that is taken each time `CG-Minimize` is called. The loss, $\mathcal{L}(\boldsymbol{\theta})$, is computed over a held-out set. `CG-Minimize`$(q_{\boldsymbol{\theta}}(\mathbf{d}), \mathbf{d}_0)$ uses conjugate gradient to minimize $q_{\boldsymbol{\theta}}(\mathbf{d})$, starting with search direction $\mathbf{d}_0$. Similar to (18), the number of CG iterations is stopped once the relative per-iteration progress made in minimizing the CG objective function falls below a certain tolerance. The `CG-Minimize` function returns a series of steps $\{\mathbf{d}_1, \mathbf{d}_2, \ldots, \mathbf{d}_N\}$ that are then used in a backtracking procedure. The parameter update, $\boldsymbol{\theta} \leftarrow \boldsymbol{\theta} + \alpha\mathbf{d}_i$, is based on an Armijo rule backtracking line search. $\beta < 1.0$ is a momentum term.

To perform distributed computation, we use a master/worker architecture in which worker processes distributed over a compute cluster perform data-parallel computation of gradients and curvature matrix-vector products and the master implements the Hessian-free optimization and coordinates the activity of the workers. All communication between the master and workers is via sockets.

Algorithm 8.1 Hessian-free optimization (after (18)).

---

initialize $\boldsymbol{\theta}$; $\mathbf{d}_0 \leftarrow \mathbf{0}$; $\lambda \leftarrow \lambda_0$; $\mathcal{L}_{\text{prev}} \leftarrow \mathcal{L}(\boldsymbol{\theta})$
while not converged **do**
    $\mathbf{g} \leftarrow \nabla\mathcal{L}(\boldsymbol{\theta})$
    Let $q_{\boldsymbol{\theta}}(\mathbf{d}) = \nabla\mathcal{L}(\boldsymbol{\theta})^T\mathbf{d} + \frac{1}{2}\mathbf{d}^T(\mathbf{G}(\boldsymbol{\theta}) + \lambda\mathbf{I})\mathbf{d}$
    $\{\mathbf{d}_1, \mathbf{d}_2, \ldots, \mathbf{d}_N\} \leftarrow \text{CG-MINIMIZE}(q_{\boldsymbol{\theta}}(\mathbf{d}), \mathbf{d}_0)$
    $\mathcal{L}_{\text{best}} \leftarrow \mathcal{L}(\boldsymbol{\theta} + \mathbf{d}_N)$
    for $i \leftarrow N-1, N-2, \ldots, 1$ **do**   % backtracking
        $\mathcal{L}_{\text{curr}} \leftarrow \mathcal{L}(\boldsymbol{\theta} + \mathbf{d}_i)$
        if $\mathcal{L}_{\text{prev}} \geq \mathcal{L}_{\text{best}} \wedge \mathcal{L}_{\text{curr}} \geq \mathcal{L}_{\text{best}}$ **then**
            $i \leftarrow i + 1$
            break
        end if
        $\mathcal{L}_{\text{best}} \leftarrow \mathcal{L}_{\text{curr}}$
    end for
    if $\mathcal{L}_{\text{prev}} < \mathcal{L}_{\text{best}}$ **then**
        $\lambda \leftarrow \frac{3}{2}\lambda$; $\mathbf{d}_0 \leftarrow \mathbf{0}$
        continue
    end if
    $\rho = (\mathcal{L}_{\text{prev}} - \mathcal{L}_{\text{best}})/q_{\boldsymbol{\theta}}(\mathbf{d}_N)$
    if $\rho < 0.25$ **then**
        $\lambda \leftarrow \frac{2}{3}\lambda$
    else if $\rho > 0.75$ **then**
        $\lambda \leftarrow \frac{3}{2}\lambda$
    end if
    $\boldsymbol{\theta} \leftarrow \boldsymbol{\theta} + \alpha\mathbf{d}_i$; $\mathbf{d}_0 \leftarrow \beta\mathbf{d}_N$; $\mathcal{L}_{\text{prev}} \leftarrow \mathcal{L}_{\text{best}}$
end while

8.5 Low-Rank Matrices

In addition to the training parallelization approaches proposed in Sections 8.3 and 8.4, we look to further improve training time for both cross-entropy and sequence training by reducing the overall number of parameters in the network through a low-rank matrix factorization. The left-hand side of Figure 8.1 shows a typical neural network architecture for speech recognition problems, namely 5 hidden layers with 1,024 hidden units per layer, and a softmax layer with 2,220 output targets. In this paper, we look to represent the last weight matrix in Layer 6, by a low-rank matrix. Specifically, let us denote the layer 6 weight by A, which is of dimension $m \times n$. If A has rank r, then there exists (32) a factorization $A = B \times C$ where B is a full-rank matrix of size $m \times r$ and C is a full-rank matrix of size $r \times n$. Thus, we want to replace matrix A by matrices B and C. Notice there is no non-linearity (i.e. sigmoid) between matrices B and C. The right-hand side of Figure 8.1 illustrates replacing the weight matrix in Layer 6, by two matrices, one of size $1,024 \times r$ and one of size $r \times 2,220$.

We can reduce the number of parameters of the system so long as the number of parameters in B (i.e., mr) and C (i.e., rn) is less than A (i.e., mn). If we would like to reduce the number of parameters in A by a fraction p, we require the following to hold.

$$mr + rn < pmn \tag{8.8}$$

solving for r in Equation 8.8 gives the following requirement for r

$$r < \frac{pmn}{m+n} \tag{8.9}$$

In Section 8.6 and 8.7 we will discuss our choice of r for specific tasks and the reduction in the number of parameters in the network.

8.6 Analysis of Optimization Techniques

8.6.1 Experimental Setup

Our initial experiments analyzing the performance of the three proposed optimization techniques are conducted on a 50 hour English Broadcast News transcription task (11) and results are reported on 101 speakers in the EARS dev04f set. An LVCSR recipe described in (31) is used to create vocal tract length normalized (VTLN) features, which are used as input features to the DNN. The DNN architecture for Broadcast News consists of a 5 layer DNN with 1,024 hidden units per layer, and a softmax layer with 2,220

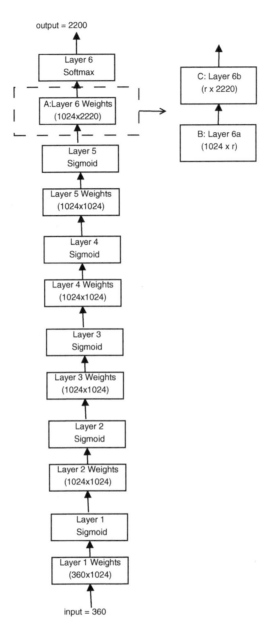

Figure 8.1: Diagram of deep neural network architecture commonly used in speech recognition.

outputs (25), as visually shown by the left hand side of Figure 8.1. For pre-training experiments, one epoch of training was done per layer for both discriminative and hybrid pre-training. For hybrid pre-training, the optimal value of α was tuned on a held-out set. For generative pre-training, multiple epochs were performed for RBM training per layer.

Following a similar recipe to (25), during fine-tuning, after one pass through the data, loss is measured on a held-out set[1] and the learning rate is annealed (i.e. reduced) by a factor of 2 if the held-out loss has not improved sufficiently over the previous iteration. Training stops after we have annealed the weights 5 times. All DNN results are reported using the cross-entropy loss function.

8.6.2 Cross Entropy Training

8.6.2.1 *WER Comparison of Pre-Training Strategies*

Table 8.1 compares the word error rate (WER) after SGD fine-tuning when generative, discriminative and hybrid pre-training is performed. Notice that the WER using discriminative pre-training is slightly worse than generative pre-training, indicating that having generalization in learning pre-trained weights is helpful. However, notice that hybrid pre-training, which combines the generalization of pre-trained weights with a discriminative objective function linked to the final cross-entropy objective function, offers a slight improvement in WER over both generative or discriminative pre-training alone.

Method	WER
Generative Pre-Training	19.6
Discriminative Pre-Training	19.7
Hybrid Pre-Training	19.5

Table 8.1: WER of Pre-Training Strategies, Broadcast News (BN)

It is natural to wonder if hybrid pre-training would produce similar results to performing generative pre-training on $x\%$ of the data and discriminative pre-training on $1 - x\%$ of the data (14). Because it is more important to generatively pre-train the lower layers and discriminatively pre-train higher layers, we explored pre-training a 5 layer DNN with a different percentage of data used for generative training per layer. A good configuration of percentage of data used for generative pre-training per layer was found to be $[80\%, 60\%, 40\%, 20\%, 0\%]$. The rest of the data per layer was used for discriminative pre-training. Using this strategy, we obtained a WER of 19.7% - worse than the hybrid pre-training WER. This shows the value of

---

1. Note that this held out set is different than `dev04f`

the joint optimization of both hybrid and generative components in hybrid pre-training.

8.6.2.2   Timing Comparison of Pre-Training Strategies

Because both discriminative and hybrid pre-training learn weights that are more closely linked to the final objective function relative to generative pre-training, this implies that fewer iterations of fine-tuning are necessary. We confirm this experimentally by showing the number of iterations and total training time of SGD fine-tuning needed for the three pre-training strategies for Broadcast News in Table 8.2. All timing experiments in this paper were run on an 8 core Intel Xeon X5570@2.93GHz CPU. Matrix/vector operations for DNN training are multi-threaded using Intel MKL-BLAS.

Notice that fewer iterations of fine-tuning are needed for both hybrid and discriminative pre-training, relative to generative pre-training. Because discriminative pre-training lacks a generative component and is even closer to the final objective function compared to hybrid pre-training, fewer fine-tuning iterations are required for discriminative pre-training. However, learning weights too greedily causes the WER with discriminative pre-training to be higher than generative pre-training, as illustrated in Table 8.1. Thus, hybrid pre-training offers the best tradeoff between WER and training time of the three pre-training strategies.

Method	Number of Iterations	Fine-Tuning Time (hrs)
Generative Pre-Training	12	42.0
Discriminative Pre-Training	9	31.5
Hybrid Pre-Training	10	35.8

Table 8.2: Fine-Tuning Time for Pre-Training Strategies, BN

8.6.2.3   Larger Mini-Batch Size

Typically when generative pre-training is performed, a mini-batch size between 128-512 is used (8) [2]. The intuition, which we will show experimentally, is the following: If the batch size is too small, parallelization of matrix-matrix multiplies on CPUs is inefficient. A batch size which is too large often makes training unstable. However, when weights are in a much better initial space,

---

2. The authors are aware that in (29) a batch size of 1,000 was used. However, the first few iterations of training were run with a batch size of 256 before increasing to 1,000.

we hypothesize that a larger batch size can be used, speeding up training time further. Figure 8.2 shows the WER as a function of batch size for both generative and hybrid pre-training methods. Note that we have not included discriminative pre-training in this analysis, since from Section 8.6.2.2 it was shown that hybrid pre-training offers the best tradeoff between WER and training time. Notice that after a batch size of 2,000, the WER of the generative pre-training method starts to rapidly increase, while with hybrid pre-training, we can have a batch size of 10,000 with no degradation in WER compared to generative PT. Even at a batch size of 20,000 the WER degradation is minimal.

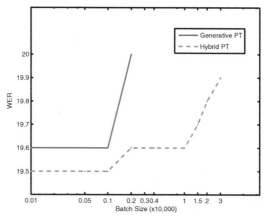

Figure 8.2: Batch Size vs. WER for Pre-Training Strategies, BN

8.6.2.4 *Parallel Stochastic Gradient Descent*

Having a large batch size implies that the gradient computations can efficiently be parallelized across worker machines. This means that the cost to compute the gradient on each machine and pass parameters around is cheaper than the cost to compute a gradient with a smaller mini-batch on one machine.

Table 8.3 shows that we can improve the fine-tuning training time by more than 1.6 using parallel SGD over serial SGD for the same batch size of 20,000. In addition, hybrid PT + parallel SGD provides a large speedup over generative pre-training. The fine-tuning training time for generative PT with a batch size of 512, a commonly used size in the literature, is roughly 24.7 hours. With hybrid PT and a batch size of 20K, the training time is roughly 9.9 hours, a 2.5x speedup over generative PT with little loss in accuracy.

Method	Fine-Tuning Training Time (hrs)
Generative PT, Serial SGD	24.7
Hybrid PT, Serial SGD	16.4
Hybrid PT, Parallel SGD	9.9 (5 workers)

Table 8.3: Comparison between Serial and Parallel SGD Fine-Tuning Training Time

8.6.2.5   Low Rank

In addition to reducing training time through parallel SGD, low-rank factorization can also be used to reduce number of parameters of the system and improve training time further. Using the low-rank factorization first involves learning the appropriate choice of r. Specifically, in the low-rank experiments, we replace the final layer matrix of size $1,024 \times 2,220$, with two matrices, one of size $1,024 \times r$ and one of size $r \times 2,220$. Table 8.4 shows the WER for different choices of the rank r and percentage reduction in parameters compared to the baseline DNN system. The number of parameters is calculated by counting the total number of trainable parameters, which includes the linear weight matrices and biases in each layer. The table shows that with a rank $r = 128$, we can achieve a 33% reduction in number of parameters, with only a slight loss in accuracy.

r = Rank	WER	Number of Params (% Reduction)
Baseline, No Rank	19.5	5.5M
512	19.6	4.9M (11%)
256	19.3	4.1M (25%)
128	19.7	3.7M (33%)
64	19.8	3.4M (38%)

Table 8.4: WER for Difference Choices of Rank r

Finally, we explore the combined speedups from both parallel SGD + low-rank factorization. We see that by including low-rank factorization, we can reduce training time by another 2 hours, leading to overall a 3-times speedup over generative PT + serial SGD. While the WER with low-rank + parallel SGD increases to 20.1%, compared to 19.5% with generative PT + serial SGD, we will show in Section 8.6.3 that after weights are re-tuned with sequence-training, this slight degradation in accuracy at the CE stage becomes negligible.

Method	Fine-Tuning Training Time (hrs)
Generative PT, Serial SGD	24.7
Hybrid PT, Parallel SGD, Full-Rank	9.9 (5 workers)
Hybrid PT, Parallel SGD, Low-Rank	7.9 (5 workers)

Table 8.5: Comparison between Serial and Parallel SGD Fine-Tuning Training Time

8.6.3 Sequence Training

In this section, we investigate potential speedups with sequence training.

8.6.3.1 *Hessian-Free*

Table 8.6 compares the WER and training time when using SGD vs. HF for sequence-training. First, notice that the minimum Bayes risk training yields relative improvements of 14–17% over cross-entropy training, with Hessian-free optimization outperforming stochastic gradient descent. The distributed Hessian-free training is also faster due to parallelization: training required 16.7 hours elapsed time with 12 workers using Hessian-free optimization, while stochastic gradient descent required 56.6 hours, more than a 3x speedup. Using more workers with HF (i.e. 48) can result in an every larger speedup, up to 5 times, as shown in (12)

model	WER	Training Time (hrs)
SA DNN cross-entropy	19.6	-
SA DNN SGD sMBR	16.9	56.6
SA DNN HF sMBR	16.2	16.7

Table 8.6: Comparison of SGD vs. HF Sequence-Trained DNN models on English broadcast news tasks.

8.6.3.2 *Low Rank*

Finally, we explore incorporating Hessian-free with low-rank. Given that $r = 128$ was best architecture for cross-entropy training, we keep this architecture for sequence training. Table 8.7 shows the performance after sequence-training for the low-rank and full-rank networks. Notice that the WER of both systems is the same, indicating that low-rank factorization does not hurt DNN performance during sequence training. In addition, even though using low-rank + parallel SGD slightly degraded the WER during CE training, this degradation disappears after sequence training.

Notice that the number of iterations for the low-rank system is significantly reduced to 12 compared to the full-rank system of 23. With a second-order hessian-free technique, the introduction of low-rank helps to further constrain the space of search directions and makes the optimization more efficient. This leads to an overall sequence training time of 4.5 hours, roughly a 3.8x speedup in training time compared to the full-rank system with a training time of 16.7 hours.

model	WER	Iters	Training Time (hrs)
SA DNN HF sMBR, no low-rank	16.2	23	16.7
SA DNN HF sMBR, low-rank	16.2	12	4.5

Table 8.7: Comparison of SGD vs. HF Sequence-Trained DNN models on English broadcast news tasks.

8.7 Performance on Larger Tasks

In this section, we explore the scalability of the three optimization methods (i.e., parallel SGD, HF and low-rank) on two larger tasks. It is too computationally intensive for us to compare WER performance with and without the proposed optimization strategies on the large tasks. Therefore, the goal of this section is to show that with the proposed optimization methods, we can still achieve similar relative improvements (i.e., between 10-20% relative) to those reported on the literature on similar tasks (12), (10),(29) with DNNs compared to state-of-the-art GMM/HMM systems. It is important to note that we have explored using low-rank for larger tasks and larger networks trained with cross-entropy. We have found that low-rank factorization continues to allow for a significant reduction in parameters compared to full-rank DNNs with no loss in accuracy, and we do not expect this trend to train with HF sequence training.

8.7.1 400 Hr Broadcast News

8.7.1.1 *Experimental Setup*

First, we explore scalability of the proposed optimization techniques on 400 hours of English Broadcast News (11). Development is done on the DARPA EARS `dev04f` set. Testing is done on the DARPA EARS `rt04` evaluation set.

The GMM system is trained using our standard recipe (31), which is briefly described below. The raw acoustic features are 19-dimensional perceptual linear predictive (PLP) features with speaker-based mean, variance, and vocal tract length normalization. Temporal context is included by splicing 9 successive frames of PLP features into supervectors, then projecting to 40 dimensions using linear discriminant analysis (LDA). The feature space is further diagonalized using a global semi-tied covariance (STC) transform (5). The GMMs are speaker-adaptively trained, with a feature-space maximum likelihood linear (FMLLR) transform estimated per speaker in training and testing. Following maximum-likelihood training of the GMMs, feature-space discriminative training (fMPE) and model-space discriminative training are done using the minimum phone error (MPE) criterion. At test time, unsupervised adaptation using regression tree MLLR is performed. The GMMs use 5,999 quinphone states and 150K diagonal-covariance Gaussians, for a total of 12.2M trainable parameters.

The DNN systems use the same FMLLR features and 5,999 quinphone states as the GMM system described above, with a 9-frame context (±4) around the current frame, and use six hidden layers each containing 1,024 sigmoidal units. Results are presented with a rank of $r = 128$, which resulted in a 49% reduction in number of parameters, from 10.7M without low-rank to 5.5M with low-rank. We refer the reader to (26) for detailed experiments regarding the choice of r for this task. FMLLR features are used instead of fMPE features because discriminative features were found to offer no advantage for DNN acoustic models (24). The DNN training begins with greedy, layerwise, hybrid pre-training and then continues with discriminative training, using the proposed parallel SGD method with a large batch size. Sequence-training is performed using Hessian-free sMBR training.

8.7.1.2 *Results*

The word error rate results are presented in Table 8.8. Prior to sMBR training, the performance of the DNN, is slightly worse than the speaker-adapted, discriminatively trained GMM. Following sMBR training, the DNN is the best model. It is 8% better than the SAT+DT GMM on `dev04f` and 9% better on `rt03`. Furthermore, we are able to achieve these results with more than 50% fewer parameters than our GMM/HMM system due to low-rank factorization.

model	dev04f	rt03
SA fMPE+MPE GMM	16.5	14.8
SA DNN cross-entropy	16.7	14.6
SA DNN HF sMBR	**15.1**	**13.4**

Table 8.8: Comparison of GMM and DNN models on Switchboard tasks.

8.7.2   300 Hr Switchboard

8.7.2.1   Experimental Setup

Second, we demonstrate scalability of the proposed optimization techniques on 300 hours of conversational American English telephony data from the Switchboard corpus. Development is done on the `Hub5'00` set, while testing is done on the `rt03` set, where we report performance separately on the Switchboard (`SWB`) and Fisher (`FSH`) portions of the set.

First, we compare speaker-independent (SI) + feature-and model-space (FMMI+BMMI) discriminatively trained GMMs to speaker-adaptive (SA) + discriminatively trained GMM models. The GMM systems are trained using the same methods described above. The speaker-adaptive results include adaptation using regression tree MLLR. The speaker-independent GMMs use 9,300 quinphone states and 370K Gaussians, for a total of 30M trainable parameters. The speaker-adaptive GMMs use 8,260 quinphone states and 372K Gaussians, for a total of 30.1M trainable parameters. The recognition vocabulary contains 30.5K words with 1.08 pronunciation variants per word. The language model is small, containing a total of 4.1M n-grams, and is an interpolated back-off 4-gram model smoothed using modified Kneser-Ney smoothing. Both the lexicon and language model are described in more detail in (2).

The DNN models and training procedure, including block size for randomization, are patterned after those in (30). We also compare speaker-independent (SI) and speaker-adaptive (SA) input features for the DNN. The SI input features are the same 40-dimensional PLP+LDA+STC features used in the SI GMM system, excluding the FMMI transform. The SA input features are also the same PLP+LDA+STC+VTLM+fMLLR features used in the SA GMM system. An input of 11 frames of context (± 5 around the current frame) is provided as input to the DNNs, which use six hidden layers each containing 2,048 hidden units, to estimate the posterior probabilities of the same 9,300 quinphone units used by the SI GMM system. Results are presented with a rank of $r = 512$, which resulted in a 32% reduction in number of parameters, from 41M parameters to 28M. Again, we refer the reader to (26) for detailed experiments regarding the choice of r for this task. The same training steps are used as in the 400-hour broadcast news task: first, hybrid pre-training with, then discriminative training with the cross-entropy criterion and parallel stochastic gradient descent, and a final sMBR optimization using distributed Hessian-free training.

8.7.2.2   Results

The word error rate results are presented in Table 8.9 for both SI and SA systems. For the SI system, after sequence training the SI DNN is 27% better than the SI GMM on rt03FSH and 13% on rt03FSH. For the SA system, after sequence training the SI DNN is 27% better than the SI GMM on rt03FSH and 13% on rt03FSH. With the low-rank factorization, the DNN models are now trained with the same number of parameters as the GMM systems.

model	rt03		Hub5'00
	FSH	SWB	SWB
SI FMMI+BMMI GMM	22.6	33.3	18.9
SI DNN cross-entropy	18.9	28.8	16.1
SI DNN HF sMBR	16.4	25.5	13.3
SA FMMI+BMMI GMM	17.6	26.3	15.1
SA DNN cross-entropy	16.8	25.4	14.2
SA DNN HF sMBR	**14.9**	**23.5**	**12.2**

Table 8.9: Comparison of GMM and DNN models on Switchboard tasks.

8.8   Conclusions and Future Work

In this paper, we introduced a variety of different optimization techniques to improve DNN training speed. Using hybrid pre-training, we are able to successfully parallelize the gradient computation, achieving a 3x speedup for cross-entropy fine-tuning time on a 50 hr English BN task. With Hessian-free optimization, sequence training can also be sped up by a factor of 3. In addition, with low-rank matrix factorization, we can reduce the number of parameters by 33% with no loss in accuracy. Finally, applying the proposed techniques, we are able to train DNNs on a 300 hr SWB task and a 400 hr English BN task, showing improvements between 9-30% relative compared to a state-of-the art GMM/HMM system while the of parameters of the

DNN is smaller than that of the GMM/HMM system. As sequence-training is the slowest part of the overall training process, in the future we would like to explore further speedup ideas related to Hessian-free sequence training.

Acknowledgements

The authors would like to thank George Saon and Stanley Chen for their contributions towards the IBM toolkit and recognizer utilized in this paper. Also, thanks to Abdel-rahman Mohamed and Vikas Sindhwani for many useful discussions related to DNNs. Furthermore, we thank James Martens for his prompt and clear answers to our questions about Hessian-free optimization. This work was supported in part by Contract No. D11PC20192 DOI/NBC under the RATS program. The views expressed are those of the author and do not reflect the official policy or position of the Department of Defense or the U.S. Government. Approved for Public Release, Distribution Unlimited.

References

1. R. Byrd, G. M. Chin, W. Neveitt, and J. Nocedal. On the use of stochastic Hessian information in unconstrained optimization. *SIAM Journal on Optimization*, 21(3):977–995, 2011.

2. S. F. Chen, B. Kingsbury, L. Mangu, D. Povey, G. Saon, H. Soltau, and G. Zweig. Advances in speech transcription at IBM under the DARPA EARS program. *IEEE Transactions on Audio, Speech, and Language Processing*, 14(5):1596–1608, 2006.

3. C. T. Chu, S. K. Kim, Y. A. Lin, Y. Y. Yu, G. Bradski, A. Ng, and K. Olukotun. Map-reduce for Machine Learning on Multicore. In *Proc. NIPS*, 2007.

4. J. Dean, G. Corrado, R. Monga, K. Chen, M. Devin, Q. Le, M. Mao, M. Ranzato, A. Senior, P. Tucker, K. Yang, and A. Y. Ng. Large Scale Distributed Deep Networks. In *NIPS*, 2012.

5. M. Gales. Semi-tied covariance matrices for hidden Markov models. *IEEE Transactions on Speech and Audio Processing*, 7(3):272–281, 1999.

6. X. Glorot and Y. Bengio. Understanding the difficulty of training deep feedforward neural networks. In *Proc. AISTATS*, pages 249–256, 2010.

7. G. Hinton, L. Deng, D. Yu, G. Dahl, A. Mohamed, N. Jaitly, A. Senior, V. Vanhoucke, P. Nguyen, T. N. Sainath, and B. Kingsbury. Deep Neural

Networks for Acoustic Modeling in Speech Recognition. *IEEE Signal Processing Magazine*, 29(6):82–97, 2012.

8. G. E. Hinton. A Practical Guide to Training Restricted Boltzmann Machines. Technical Report 2010-003, Machine Learning Group, University of Toronto, 2010.

9. G. E. Hinton, S. Osindero, and Y. Teh. A Fast Learning Algorithm for Deep Belief Nets. *Neural Computation*, 18:1527–1554, 2006.

10. N. Jaitly, P. Nguyen, A. W. Senior, and V. Vanhoucke. Application Of Pretrained Deep Neural Networks To Large Vocabulary Speech Recognition. In *Proc. Interspeech*, 2012.

11. B. Kingsbury. Lattice-Based Optimization of Sequence Classification Criteria for Neural-Network Acoustic Modeling. In *Proc. ICASSP*, 2009.

12. B. Kingsbury, T. N. Sainath, and H. Soltau. Scalable Minimum Bayes Risk Training of Deep Neural Network Acoustic Models Using Distributed Hessian-free Optimization. In *Proc. Interspeech*, 2012.

13. H. Larochelle and Y. Bengio. Classification Using Discriminative Restricted Boltzmann Machines. In *Proc. ICML*, 2008.

14. H. Larochelle, Y. Bengio, J. Louradour, and P. Lamblin. Exploring Strategies for Training Deep Neural Networks. *Journal of Machine Learning Research*, 1:1–40, 2009.

15. Q. Le, J. Ngiam, A. Coates, A. Lahiri, B. Pronchnow, and A. Ng. On Optimization Methods for Deep Learning. In *Proc. ICML*, 2011.

16. Y. LeCun and Y. Bengio. Convolutional Networks for Images, Speech, and Time-series. In *The Handbook of Brain Theory and Neural Networks*. MIT Press, 1995.

17. Y. LeCun, J. S. Denker, S. Solla, R. E. Howard, and L. D. Jackel. Optimal Brain Damage. In *Advances in Neural Information Processing Systems 2*, 1990.

18. J. Martens. Deep learning via Hessian-free optimization. In *Proc. Intl. Conf. on Machine Learning (ICML)*, 2010.

19. A. Mohamed, G. Hinton, and G. Penn. Understanding how Deep Belief Networks Perform Acoustic Modeling. In *Proc. ICASSP*, 2012.

20. B. A. Pearlmutter. Fast exact multiplication by the Hessian. *Neural Computation*, 6(1):147–160, 1994.

21. R. Raina, A. Madhavan, and A. Y. Ng. Large-scale Deep Unsupervised Learning Using Graphics Processors. In *Proc. ICML*, 2009.

22. B. Recht and C. Re. Parallel Stochastic Gradient Algorithms for Large-Scale Matrix Completion. *Optimization Online*, 2011.

23. T. N. Sainath, B. Kingsbury, A. Mohamed, and B. Ramabhadran. Convolutional Neural Networks for Large Vocabulary Speech Recognition. In *submitted to Proc. ICASSP*, 2013.

24. T. N. Sainath, B. Kingsbury, and B. Ramabhadran. Improvements in Using Deep Belief Networks for Large Vocabulary Continuous Speech Recognition. Technical report, IBM, 2012.

25. T. N. Sainath, B. Kingsbury, B. Ramabhadran, P. Fousek, P. Novak, and A. Mohamed. Making Deep Belief Networks Effective for Large Vocabulary Continuous Speech Recognition. In *Proc. ASRU*, 2011.

26. T. N. Sainath, B. Kingsbury, V. Sindhwani, E. Arisoy, and B. Ramabhadran. Low-Rank Matrix Factorization for Deep Belief Network Training. In *submitted to Proc. ICASSP*, 2013.

27. T. N. Sainath and B. Kingsbury and B. Ramabhadran. Improving Training Time of Deep Belief Networks Through Hybrid Pre-Training And Larger Batch Sizes. In *to appear in Proc. NIPS Workshop on Log-linear Models*, 2012.

28. N. N. Schraudolph. Fast curvature matrix-vector products for second-order gradient descent. *Neural Computation*, 14:1723–1738, 2004.

29. F. Seide, G. Li, X. Chen, and D. Yu. Feature engineering in context-dependent deep neural networks for conversational speech transcription. In *Proc. ASRU*, pages 24–29, 2011.

30. F. Seide, G. Li, and D. Yu. Conversational Speech Transcription Using Context-Dependent Deep Neural Networks. In *Proc. Interspeech*, 2011.

31. H. Soltau, G. Saon, and B. Kingsbury. The IBM Attila speech recognition toolkit. In *Proc. IEEE Workshop on Spoken Language Technology*, pages 97–102, 2010.

32. G. Strang. *Introduction to Linear Algebra*. Wellesley-Cambridge Press, 4th edition, 2009.

33. K. Vesely, L. Burget, and F. Grezl. Parallel Training of Neural Networks for Speech Recognition. In *Proc. Interspeech*, 2010.

34. O. Vinyals and D. Povey. Krylov subspace descent for deep learning. In *Proc. NIPS Workshop on Optimization and Hierarchical Learning*, 2011.

35. D. Yu, L. Deng, and G. E. Dahl. Roles of Pre-Training and Fine-Tuning in Context-Dependent DBN-HMMs for Real-World Speech Recognition. In *NIPS Workshop on Deep Learning and Unsupervised Feature Learning*, 2010.

36. D. Yu, F. Seide, G. Li, and L. Deng. Exploiting Sparseness In Deep Neural Networks For Large Vocabulary Speech Recognition. In *Proc. ICASSP*, 2012.

37. M. Yuan, A. Ekici, Z. Lu, and R. Monteiro. Dimension Reduction and Coefficient Estimation in Multivariate Linear Regression. *J. R. Statist. Soc. B.*, 69(3):329–346, 2007.

38. M. A. Zinkevich, M. Weimer, A. Smola, and L. Li. Parallelized Stochastic Gradient Descent. In *Proc. NIPS*, 2010.

Neural Information Processing series

Michael I. Jordan and Thomas Diettrich, Editors

Advances in Large Margin Classifiers, Alexander J. Smola, Peter L. Bartlett, Bernhard Scholkopf, and Dale Schuurmans, eds., 2000

Advanced Mean Field Methods: Theory and Practice, Manfred Opper, and David Saad, eds., 2001

Probabilistic Models of the Brain: Perception and Neural Function, Rajesh P. N. Rao, Bruno A. Olshausen, and Michael S. Lewicki, eds., 2002

Exploratory Analysis and Data Modeling in Functional Neuroimaging, Friedrich T. Sommer, and Andrzej Wichert, eds., 2003

Advances in Minimum Description Length: Theory and Applications, Peter D. Grunwald, In Jae Myung, and Mark A. Pitt, eds., 2005

Nearest-Neighbor Methods in Learning and Vision: Theory and Practice, Gregory Shakhnarovich, Piotr Indyk, and Trevor Darrell, eds., 2006

New Directions in Statistical Signal Processing: From Systems to Brains, Simon Haykin, Jose C. Príncipe, Terrence J. Sejnowski, and John McWhirter, eds., 2007

Predicting Structured Data, Gokhan Bakˇr, Thomas Hofmann, Bernhard Scholkopf, Alexander J. Smola, Ben Taskar, and S. V. N. Vishwanathan, eds., 2007

Toward Brain-Computer Interfacing, Guido Dornhege, Jose del R. Millan, Thilo Hinterberger, Dennis J. McFarland, and Klaus-Robert Müller, eds., 2007

Large Scale Kernel Machines, Leon Bottou, Olivier Chapelle, Denis De-Coste, and Jason Weston, eds., 2007

Learning Machine Translation, Cyril Goutte, Nicola Cancedda, Marc Dymetman, and George Foster, eds., 2009

Dataset Shift in Machine Learning, Joaquin Quiñonero-Candela, Masashi Sugiyama, Anton Schwaighofer, and Neil D. Lawrence, eds., 2009

Optimization for Machine Learning, Suvrit Sra, Sebastian Nowozin, and Stephen J. Wright, eds., 2012

Practical Applications of Sparse Modeling, Irina Rish, Guillermo A. Cecchi, Aurelie Lozano, and Alexandru Niculescu-Mizil, eds., 2014

Advanced Structured Prediction, Sebastian Nowozin, Peter V. Gehler, Jeremy Jancsary, and Christoph H. Lampert, eds., 2015

Perturbations, Optimization, and Statistics, Tamir Hazan, George Papandreou, and Daniel Tarlow, eds., 2016

Log-Linear Models, Extensions and Applications, Aleksandr Aravkin, Anna Choromanska, Li Deng, Georg Heigold, Tony Jebara, Dimitri Kanevsky, and Stephen J. Wright, eds., 2019